Why Carbon Fuels Will Dominate The 21st Century's Global Energy Economy

PETER R. ODELL

Professor Emeritus of International Energy Studies

Erasmus University

Rotterdam

© 2004 Peter Odell

Peter Odell has asserted his right under the Copyright, Designs and Patents Act 1998 to be identified as the author of this work.

All rights reserved. Except for brief quotations in review, this book, or any part thereof, may not be reproduced, stored in or introduced into a retrieval system or transmitted, in any form or by any means, electronic, mechanical, photocopying, recording or otherwise, without the prior written permission of the author or publisher.

ISBN 0 906522 22 6
Multi-Science Publishing Co. Ltd.
5 Wates Way, Brentwood, Essex CM15 9TB, UK

*To the librarian and her colleagues in
the Library and Information Service
at the Institute of Petroleum (now the
Energy Institute) in London.*

ACKNOWLEDGEMENTS

This book is derived, in part, from an earlier shorter study, *Fossil Fuel Resources in the 21st Century*, commissioned from me in 1998 by the Planning and Economic Studies Section of the Division of Nuclear Power and the Fuel Cycle in the International Atomic Energy Agency in Vienna.

Forty years prior to that commission I had been introduced to energy sector forecasting in the Economics Division of Shell International. My interest in this was subsequently developed in my tenure at the London School of Economics and Erasmus University Rotterdam. At the latter institution my initial efforts to interpret the long-term future of oil and gas were dependent on the contributions of Dr. K. E. Rosing and on a succession of research students.

During the late 70s to the early 90s an ongoing involvement in the International Energy Workshop at the International Institute for Applied Systems Analysis in Vienna and my participation in many conferences of the International Association for Energy Economics and in seminars at the Canadian Energy Research Institute in Calgary and the International Energy Agency in Paris etc provided stimulating environments in which further to refine ideas on the long-term future of the global energy economy.

This book presents the culmination of my efforts to draw some conclusions on the prospects for energy in the 21st century arising from the inputs provided by many colleagues at the aforementioned institutions. I may yet live long enough to regret the foolhardiness of the attempt; and/or to persuade others to define more robust forecasts on one of the most important variables for humankind's prospects.

Finally, I am indebted to Dr. Helga Graham's willingness to appraise the study and for her many insightful comments on its contents and the consistency of its arguments. Needless to say, errors in, and omissions from, the study – together with the predictions which prove to be faulty – are my sole responsibility.

Peter R. Odell
London, UK.

peter@odell.u-net.com

CONTENTS

List of Figures	vii
List of Tables	viii
Preface and Overview	**xi**
Preface	xi
Overview	xix
The Demand for Energy	xix
Global Energy Production	xx
The Relative Decline Of Coal	xx
Hydrocarbons' Role	xxi
The Declining Importance Of Oil	xxi
Natural Gas – The Fuel Of The 21st Century	xxii
Oil and Gas as Renewable Resources	xxiii
Chapter 1: The Global Energy Economy in 2000:	
and its Prospects for the 21st Century	**1**
Historical background	1
Prospects for energy demand	9
Carbon fuel resources	12
References	16
Chapter 2: Global Coal Resources and Production Prospects	**19**
Long-term availability	19
Environmental constraints	22
Regional considerations	23
Western and Central Europe	26
South East Asia and the Western Pacific Rim	27
The Rest of the World	29
Possible new developments	30
References	32
Chapter 3: Oil's Long-term Future: 85% yet to be Exploited	**35**
The reserves' discovery and appreciation process	35
What oil crisis?	41
Ultimate conventional oil resources' depletion,	
1940–2140	45
Non-conventional oil enters the market	49

Environmental constraints on oil's production, transport and use	55
Regional issues in oil supply prospects	60
Possible new developments	65
References	66

Chapter 4: Natural Gas – The Prime Energy Source for the 21st Century — 71

Resource abundance	71
Production potential overall	76
Natural gas' environmentally friendly characteristics	80
Regional Gas markets	82
North America	83
Central and South America	85
Europe (excluding the FSU)	85
The Former Soviet Union	87
The Middle East	89
Africa	91
Asia Pacific	93
Significant new developments for gas' expansion	94
References	97

Chapter 5: Trends in Production Costs and Prices — 101

Introduction	101
Supply costs for carbon fuels	102
Prices to 2040	104
Mid-century prices	105
Prices post-2060	107
An overview of developments	107
References	108

Chapter 6: Oil and Gas as Renewable Resources? — 111

Biogenic carbon energies' limitations	111
The abiogenic theory of the origins of petroleum	112
The theory ignored to date: is this about the change?	120
References	122

Postscript	123
Bibliography	129
Index	147

LIST OF FIGURES

Figure 1.1:	Trends in the evolution of world energy use, 1860–2000	2
Figure 1.2:	Relationship between energy use and economic development over time	3
Figure 1.3:	The price of Saudi Arabian/Dubai light crude oil, 1950–2000	6
Figure 1.4:	Trends in the evolution of energy supplies by source and per decade in the 21st century	11
Figure 1.5:	Cumulative supplies of energy, by source, in the 21st century	15
Figure 3.1:	BP's 1979 view of oil depletion	37
Figure 3.2:	The appreciation of proven reserves of conventional oil compared with cumulative demand, 1945–2000	40
Figure 3.3:	Assessments of total world initial oil reserves over the period, 1940–2000	47
Figure 3.4:	Production curves for conventional and non-conventional oil, 1940–2140	49
Figure 3.5:	The complementary relationship of conventional and non-conventional oil production, 1940–2140	49
Figure 3.6:	The changing contributions of oil and gas supplies by decade in the 21st century	54
Figure 3.7:	Cumulative oil and gas Production in the 21st Century	55
Figure 4.1:	Production curves for conventional and non-conventional gas, 1940–2140	78
Figure 4.2:	The complementary relationship of conventional and non-conventional gas production, 1940–2140	78
Figure 4.3:	21st Century oil and gas supplies by decade	81
Figure 4.4:	Europe's emerging continental gas supply system	88
Figure 4.5:	Boundary claims in the South China Sea	95

LIST OF TABLES

Table 2.1:	Assessments of world coal resources, 1936–1996	20
Table 2.2:	Proven coal reserves, 1936–2001	21
Table 2.3:	Rank ordering of leading countries' coal reserves compared with shares of world production, 2000	24
Table 2.4:	Rank ordering of leading coal exporters, 2000	24
Table 2.5:	Rank ordering of leading coal importers, 2000	25
Table 2.6:	Rank ordering of countries at least 20% coal-dependent in 2000	30
Table 3.1:	Proven reserves, oil production, gross and net growth in oil reserves and reserves-to-production ratios, 1971–2000	38
Table 3.2:	The changing contributions of oil to the total supply of hydrocarbons, 2000–2050 and 2100	53
Table 3.3:	The cumulative contributions of oil and natural gas to the energy supply in the 21st century	56
Table 3.4:	Rank-ordering of the top-10 oil producing and consuming countries in 2000	61
Table 4.1:	World conventional gas reserves and resources by region and percentage depletion by 2000	74
Table 4.2:	World non-conventional gas resources by region	76
Table 4.3:	Contribution of natural gas to total energy and carbon fuel use by region, 2000	83

AUTHOR'S NOTE

The following terms have been used:

- Tons always means metric tons

- Gtoe and mtoe refer to gigatons (10^9) and million (10^6) tons of oil equivalent, respectively

- One thousand million (10^9) is designated as a billion (B) and one million million (10^{12}) as a trillion (T) with respect to barrels of oil and cubic metres of gas

FREQUENTLY USED ACRONYMS AND ABBREVIATIONS

AAPG	American Association of Petroleum Geologists
BP	British Petroleum plc
EC	European Commission
EIU	Economist Intelligence Unit
EU	European Union
IEA	International Energy Agency
IGU	International Gas Union
IIASA	International Institute for Applied Systems Analysis
IPCC	Inter-governmental Panel on Climate Change
OAPEC	Organisation of Arab Petroleum Exporting Countries
OECD	Organisation for Economic Cooperation and Development
OGJ	Oil and Gas Journal
OPEC	Organisation of Petroleum Exporting Countries
RIIA	Royal Institute of International Affairs
Shell IPC	Shell International Petroleum Company Ltd
WEC	World Energy Council
WPC	World Petroleum Congress

Preface and Overview

A. Preface

Realism over the critical issues of energy supply and use in the 21st century's economies and societies has become a very scarce commodity. This has emerged from a combination of three widely presented, but controversial, hypotheses, viz. first, that there is an inherent scarcity in the world's endowment of energy resources (Campbell, 2003; Heinberg, 2003); second, that a rapid onset of global warming and climatic change will be a consequence of anthropogenically derived CO_2 emissions into the atmosphere (Bossel, 1998; Meadows, 2002); and third, that a set of geopolitical constraints will inevitably inhibit the production of, and trade in, energy (Claes, 2001; Mitchell J. et al, 2001). Individually, each of these beliefs implies a relatively near-future requirement for moderating the current degree of dependence on carbon fuels*; while, collectively, the three concerns not only enhance, but also accelerate the perceived need for a comprehensive switch to the use of alternative energy sources. The objective of this study is, however, to demonstrate that the moves to economies and societies wholly or largely free from dependence on carbon energy are, in the real world, incapable of being achieved.

The greater part of the book is thus dedicated to showing that, for most of the 21st century, energy demand limitations will be so significant that little or no pressure will be brought to bear on the

---

* These are often referred to as "fossil fuels" based on the theory of a biogenic origin for oil, coal and natural gas. As this is by no means universally accepted – as shown in Chapter 6 of this book – the term carbon energy is used throughout the book to define the world's three principal energy resources.

relatively plentiful and profitable-to-produce flows of coal, oil and natural gas. Indeed, continuity in the slowly increasing supply of carbon energy – based on a modest depletion of the world's generous coal resource base and on the exploitation of about three-quarters of the world's currently conservatively-estimated remaining 5000 billion barrels of oil – will be achieved; albeit in part, as a result of the accelerating substitution of coal and oil products by natural gas, so creating a successful evolution of the markets for carbon fuels for at least the first half of the century. Thereafter, plentiful natural gas resources – partly conventional, but more significantly, unconventional – can readily sustain most of the total potential energy supply required until the very last decade of the 21st century. In so doing, the world's natural gas industry will, by 2100, be more than five times its size in 2000.

Over the 21st century as a whole, a total of some 1660 Gigatons (= 1660 x 10^9 tons) oil equivalent of carbon energy will be produced and used, compared with a cumulative total in the 20th century of just under 500 Gigatons. This more-than-three-fold increase in the use of carbon energy in the 21st century reflects not only the bountiful nature of the world's endowment of carbon energy fuels, but also the willingness of the nations which are rich in coal, oil and/or natural gas to accept the depletion of their "natural" resources, in return for the economic growth it generates for the countries concerned and the rising incomes it secures for their populations.

It also indicates the managerial and technological achievements which can be anticipated through the multitude of global regional and local entities responsible for the extraction, the transportation and the processing of the world's energy resources. The fundamental mutuality of the interests of the very many parties already involved in such activities – albeit with temporary disturbances between them arising from economic and/or political difficulties (as over the past 100 years) – will virtually ensure supply continuity at the levels required by demand developments. In this set of defined circumstances for the exploitation of carbon energies, the concept of "resource wars" (Klare, 2002; Kleveman, 2003) becomes invalid, as such phenomena are likely only in the context of a terminal scarcity of coal, oil and/or natural gas. This study demonstrates that such scarcity is excludable, except on a local or regional scale from time to time, for the 21st century.

Neither is the carbon-energy production industry a serious or even relevant phenomenon with respect to the issues of global warming and climate change: except under the close-to-unthinkable circumstances of very large scale and long-continuing releases of methane (natural gas) to the atmosphere from the production and transportation infrastructure of the industry. This could occur only in the context of the generally expected markets for gas failing to materialise, so that the companies and other entities involved had neither the will, nor any commercial motivation, to inhibit such a development.

Ironically, the only possible cause of such an occurrence would be a rapid and low-cost expansion of renewable energy sources so that the 'bottom' traumatically dropped out of the natural gas markets. In reality, neither the time involved in constructing renewable energy production plants (viz. windmills, solar power installations, tidal or wave power driven generators, biomass fuelled electricity production etc), nor the inability of such plants to produce alternative (renewable) power competitively offer a challenge to electricity made from gas – or, indeed, from coal or oil. It is, indeed, these negative attributes of most renewable energy production which make its expansion at a rate whereby renewables could meet even the *incremental* demands for energy in the 21st century quite impossible.

This inability has already been effectively demonstrated in the world's richest and most technologically orientated countries since 1990, the base year from which the Kyoto Protocol requires their use of carbon fuels and, hence, their volumes of CO_2 emissions, to be reduced. Instead, these countries' collective use of 3500 million tons of oil equivalent in 1990 (from a melange of oil, gas and coal) increased to 4100mtoe in 2002. In marked contrast with this 600mtoe rise in carbon energy use, their use of renewables increased by only 200Gtoe. Of this, moreover, 120 mtoe was accounted for by nuclear power – a pseudo-renewable energy source. But the supply of nuclear electricity has now just about peaked, given that the small number of new stations currently under construction or planned will fail to replace the output of the stations which are scheduled for decommissioning in the short and medium terms (Grimston and Beck, 2002).

To date, moreover, the status of other non-renewable energy producers (except for hydro-power) remains that of an "infant industry". An industry, that is, that is incapable of sustaining growth without either a continuing input of state subsidies to reduce production costs or the willingness of consumers to pay a premium price – over that for carbon energy – for so-called "green" energy. Thus, even for the world's already "well-energized" economies and societies – still more than 85% dependent on carbon fuels and with half of the remaining 15% derived from a nuclear power industry in decline –

there are no realistic prospects even for their incremental demand for energy to be totally met from renewables, let alone for them being capable of substituting the countries' existing use of carbon energy!

Unless, that is, the governments of these countries stipulate *and* require an energy market which is so transformed. The supply disruptions and the populist protests against the burdens of both additional capital and running costs which would consequently emerge, make such a radical policy well-nigh impossible. Such a change is thus an extremely unlikely development in the overall energy economies of these countries, in general; while, in particular, it is an impossible change for the near-exclusively carbon-fuelled transportation sector of their economies. Over 50% of global oil use and about 22% of energy use are already concentrated in this single sector – and the percentages are still increasing (Mitchell et al, 2001).

Thus, in spite of the rich world's countries' so-called Kyoto Treaty "commitments" to reduce CO_2 emissions (assuming the Treaty is eventually ratified), future progress towards the achievements required remains highly improbable (IEA, 2002a; Jean-Baptiste and Ducroux, 2003). At best, progress in reducing emissions will be slow until 2020, but with some hope thereafter of more rapid progress. This will most likely be associated with an increasingly large-scale sequestration of CO_2 captured from the combustion of carbon fuels. The technological developments, the effective management and the falling real costs of sequestration will make this a more acceptable and a financially less costly way of achieving emissions' reduction targets than that which could be achieved from constraints on carbon fuels' use, given the consequential adverse effects of such constraints on economic growth and on public opinion. Offsetting the costs of sequestration will, moreover, directly motivate attempts to enhance energy use efficiency and will also stimulate changes in economic and societal structures designed to reduce energy requirements. Albeit with a delay of a decade or more, one may also reasonably expect similar developments in the until-recently centrally planned economies of the former Soviet Union and Eastern Europe.

These two sets of countries – with only one-fifth of the world's present population – currently account for almost two-thirds of global energy use. But, a combination of their relatively low rates of population growth and their ability to achieve higher efficiencies in energy use will continue to reduce their share of the world's use of energy – and, in due course, could eventually lead to the stabilisation of their CO_2 emissions. Such progress will, moreover, be accompanied by the sequestration of their CO_2 emissions on an increasing scale.

Under these emerging circumstances the world's developing countries – already with 80% of the world's population and with the percentage still growing – will play a rising relative role in both global energy use and in CO_2 emissions. Indeed, as in most other attributes related to the process of development, these countries 'need' to use increasing amounts of energy, which, on an average per capita basis, is only one-eighth of that in the rich counties of the world. This "natural" phenomenon of rising per capita energy inputs to economic and social advancement is, however, possible, in by far the greater part, only by the countries' increasing their use of low-cost carbon fuels. The alternative renewable sources of energy are – to an even greater extent than in the developed countries – simply too-high cost, except in niche markets largely unrelated to industrialisation, urbanisation and motorisation (Anderson, 2001).

Thus, the future global energy needs of the developing world will inevitably be low-cost coal, oil and natural gas – albeit increasingly used at the higher efficiencies already achieved in the rich world – in preference to the generally higher capital-cost renewables such as are now under development in the OECD countries with subsidies from both international organisations, viz. the International Energy Agency and the European Commission, and from national governments (European Environmental Agency, 2002; European Commission, 2003; IEA, 1997). Given that higher per capita incomes, enhanced standards of social welfare and significant spatial mobility for the populations in the countries early to industrialise and urbanise steadily and cumulatively accrued over the many decades from the late 18th century to the present day, largely as a result of the continuing availability of, and access to, energy sources at low prices, then similar opportunities that are now emerging for the world's other countries and their rapidly growing populations cannot be denied to them. Unless, that is, the denial is made in the context of subsidies from the 'north' to the 'south' which fully offset the higher costs – both financial and temporal – which the production and use of large volumes of renewable energies inevitably involved. This is, to put it bluntly, highly unlikely.

Meanwhile, demographic trends will locate the overwhelming percentage of the close-to-three billion more people expected to inhabit the earth by 2050 in the developing world. All of these should expect, as a matter of course, to have access to enhanced supplies of energy, in general, and to electricity, in particular: together with the estimated present close-to-two billion people in the developing world who are still without access to domestic electricity (Anderson, 2001; IEA, 2002c; 2003c).

This necessary – rather than simply desirable – completion of access to electricity for all the world's householders is a more realistic and positive form of sustainability than that which OECD policymakers present as their top priority, viz. the achievement of global sustainability through the containment – and the relatively near-future reduction – of CO_2 emissions to the globe's atmosphere so that their hypothesised fears for global warming and climatic change can be eliminated. Quite apart from continuing doubts over these hypothesised links (Gerholm, 1999; Bradley, 2003) and the potential developments in the relatively near future of technologies which can significantly reduce and eventually eliminate the growth in atmospheric CO_2 at a cost well below that of changing from carbon fuels to renewables (Williams, 1998; Freund, 2002), there can be neither economic nor ethical justification for actions which delay or even obstruct the poor world's needs for sufficiency of energy to secure development and enhanced living standards. Most of the people concerned live in, or will be born into, countries that can only achieve such improvements through the exploitation of coal, oil and/or natural gas (Odell, 1984 and 1990; Greenpeace, 1997; IEA, 2003a).

Thus, instead of the high profile demands of the proponents of the Kyoto Treaty, through heavy-handed and urgent pressures to secure the substitution of carbon fuels around the world by the direct and indirect use of solar power, this study firmly relegates the enhancement of the presently low percentage contribution of renewable energies to the total energy supply to the second half of the 21st century. Renewables, it is predicted (see Figure 1.4) will, by 2050, still contribute no more than 20% of total global energy supply (compared with a little over 10% in 2000 – including nuclear power, but excluding non-commercial collected biomass in the world's poorer countries). Such a 100% enhancement of renewable energy's relative importance by the mid-21st century can be defined as an organic growth rate, rather than one forced through by policies which require fundamental societal changes and the denial of the use of low-cost carbon fuels. The additional use of renewables will, thus, be concentrated mainly in those countries which are severely constrained, or even completely devoid, of exploitable indigenous sources of carbon energy, so that renewables' exploitation is an imperative for their advancement. Examples of such countries are Belgium, the Czech Republic, France, Sweden, Germany and Belarus in Europe; Chile,

Paraguay and Uruguay in the Americas; Japan, the Philippines, South Korea and Turkey in Asia and a number of countries in sub-Saharan Africa. For these countries, near-future enhanced security of energy supply, rather than concern for long-term climate change possibilities, is of the essence (Thomas and Ramberg, 1990; Randall, 2002).

Post-2050, however, following the peak of global oil production and in the context of a possible reduction in the growth rate of the natural gas industry, there will be a market-orientated widening of interest in renewables, especially in the rising number of countries in which indigenous carbon fuels become relatively scarce and more expensive. As a result, renewables could, by 2080, account for over 30% of global energy use; and for over 40% by 2100. Nevertheless, even then carbon fuels will collectively still be the more important component in energy supply. But by then the world will have become emphatically marked by significant regional and country-by-country variations in energy-use patterns as a function of highly significant geographical contrasts in the availability of carbon fuels and of renewables. Cumulatively over the century, renewables will have supplied just under 30% of the total energy used (see Figure 1.5).

This predicated division of the 21st century global energy market between carbon fuels and renewables, in a world in which the population has stabilised at about nine billion of which the overwhelming majority are linked into gas and electricity systems, represents in large part a continuation of the organisation of the energy sector in a way which has already become the norm in the world's richer countries. With the indicated near four-fold increase in total annual energy use over the century (see Figure 1.4), but with an increase of only about 50% in the world's population, average per capita use of energy will increase by 2.6 times. This generalised statistic does, however, conceal a wide range of changing per capita energy use variations across the world, viz. from close to zero or even negative increases in the world's already energy intensive countries (viz. the current OECD countries minus Mexico, plus the formerly centrally planned economies of the 1950–1990 Soviet bloc), to multi-fold increases per head in the populations of today's poorest under-developed countries. In almost all cases, however, the efficiency of energy use in terms of the GDP generated by a unit of energy input will certainly have increased: and will also have reduced the rate of growth in greenhouse gas emissions.

Nevertheless, the indicated inevitability of continuing increases in the production and use of carbon energy sources seem likely to cause consternation in the ranks of the believers in a causal link between CO_2 emissions from the combustion of such fuels and global warming/climate change. For those pessimists who visualise only adverse results from such developments (Bradley, 2003), this study's conclusions incorporate only two "saving graces"; first, that only a two and a half-fold increase in carbon energy use is indicated, compared with a more than five-fold increase predicated by the IPCC's basic scenario (IPCC, 2001); and, second, that the forecast strong increase for natural gas in the mix of carbon energy sources serves to reduce CO_2 emissions by about 10% from what they would have been, had the year 2000 division of the carbon energy market remained the same in the 21st century.

The conclusions of this study for CO_2 emissions in 2100 are thus indicated at 2.25 times their 2000 level; though relatively modest, this remains self-evidently an "unsustainable" proposition for the global warming lobbyists (Bossel, 1998; Bartsch and Müller, 2000; Hoffman, 2001; European Commission, 2002; Meadows, 2002; Jean-Baptiste, 2003). Given a widely-held acceptance of the latter's claims and warnings, then the only way out of the impasse created by the inevitability and, as argued here, the economic and social desirability of increased carbon energy use, lies in an immediate start to the implementation of measures to sequestrate the CO_2 produced by combustion (Freund, 2002). The costs of this procedure will, given the difficulties, ranging from the technological to the political, which remain to be resolved (Maritus, 2003; Pachin, 2003; Torp, 2001), necessarily, albeit modestly, increase the costs – and hence the price – of carbon energy: but this price-impact on users will have a positive feed-back effect on the rate of enhancement in the efficiency of carbon energy use and will thus, in due course, serve to reduce the rate of growth in the demand for energy; and thus of CO_2 emissions.

Under these circumstances both the volumes of carbon energy required over the century could turn out to be significantly lower than this study suggest, while its share of the total energy market would be below the 70% calculated. Nevertheless, the 21st century energy economy will remain dominated by carbon fuels, albeit with much of their supply and use orientated to environmentally more friendly modes of production and transformation: most notably in natural gas' potential use for the production of hydrogen, initially for inputs to fuel cells, costs permitting (Anthrop, 2003) and, later, for the direct use of the hydrogen in both static and mobile outlets (Griffiths, 2001; Hoffman, 2001; World Petroleum Congress, 2002; European Commission, 2003; IEA, 2003b; Rifkin, 2003).

B. Overview
Although regularly recurring fears of the impending scarcity of non-renewable energy have all proved groundless, the issue of the potential availability of coal, oil and natural gas seems to remain one of concern, reflecting the world's continuing overwhelming dependence on them for its energy needs. This concern, however, is misplaced, given recent important changes on both the supply and demand sides.

The Demand for Energy
Since 1973 the rate of increase in the global use of energy has fallen back to well under its 1860–1945 long-term trend of about 2.2% a year, compared with ±5% a year from 1945 to 1973. The probability of a return in the future to the latter much higher annual rate of growth is now close to zero, given that it reflected the temporary combination of a set of conditions which cannot recur. Thus, a realistic consideration of long-term energy supply requirements now has to be orientated to a modest rate of growth in the use of energy, even before taking into account the possible impact of environmental concerns on non-renewable energy consumption or the now slowly accelerating pace of growth in renewable energies, emanating from evolving technologies of direct and indirect solar energy production.

Therefore, the long-term future availability of carbon energy sources can be considered relative to a growth of 2% a year as the highest likely requirement. Moreover, the year 2000 base from which we must now consider demand for coal, oil and gas in the 21st century is considerably lower than that which was indicated by earlier conventional forecasts. Thus, only about 26 billion barrels of oil were used in 2000 compared with the 1972 forecast of almost 140 billion. Likewise, the cumulative use of oil from 1971 to 2000 was less than 700 billion barrels, rather than the 1,750 billion anticipated. Consequently much more oil which was already discovered by 1971 remains to be used in the 21st century. Similarly for coal: a 1980 World Coal Study forecast that annual use would increase from the then current level of 3000 million tons to 10,000 million by 2000. Instead there was a mere 10% growth to 3300 million tons. Meanwhile, proven reserves of coal have increased by 50%. It is hardly surprising that views on the future availability of carbon energy and the supply potential associated with them can now be so much more relaxed than they were 30 years ago.

Global Energy Production

Chapter 1 sets out the predicted trend in the annual production of energy supplies by source during the 21st century – given an energy demand growth that is sustained at ±2% a year until mid-century, and thereafter, at a steadily declining rate of growth per decade through to 2100, in response to the cessation of population growth and an increasingly efficient use of energy. Non-renewable energy sources remain overwhelmingly dominant until 2080. Alternatives to carbon fuels will not exceed 25% of total energy supply until the late 2060s and, even by the end of the century, they will still achieve only just over 40% of supply. Cumulatively over the century, renewable energy sources will have supplied under 30% of the total energy used. Of this over 50% will be supplied in the last three decades. Unless and until the governments and peoples of the world not only accept the desirability of a much faster switch to renewable energy, but also take the necessary steps to implement the change, then global energy use in the 21st century will remain heavily orientated towards a combination of coal, oil and natural gas. To date, there are very few serious signs of these required actions for change being met.

The Relative Decline Of Coal

Nevertheless, for both economic and environmental reasons, the relative importance of the three carbon fuels will show marked changes from the contemporary situation. Moreover, the current widely accepted conventional wisdom of inevitably constrained hydrocarbon energy supplies can be shown to be based on misconceptions concerning potentially available resources. As shown in Chapter 2, coal resources are usually indicated as being an order of magnitude greater than those of oil or gas and thus implicitly, or sometimes even explicitly, it is assumed that coal must become more important than oil and gas and, eventually, the dominant component in the 21st century's carbon fuel supplies. Given coal's general lack of acceptability and, even more so, its highly geographically concentrated pattern of production (over 80% from only ten countries), this prospect was never a realistic one, but even that small likelihood has now been undermined by a combination of local, regional and global environmental concerns over coal production and use. Thus, coal's share of global carbon fuel production will fail to increase over the whole of the century during which it will contribute only just over 25% of cumulative non-renewable energy use.

Hydrocarbons' Role
It is thus oil and gas between them which will continue to supply most of the world's non-renewable energy supply until the end of the century and consequently show a near-three-fold increase in their annual supply by 2100. The relative contributions of oil and gas to the total hydrocarbon supply will, however, change radically (as discussed in Chapters 3 and 4). This change is, in part, a function of possible long-term constraints on oil supplies, and, in part, a reflection of the inherent advantages of natural gas in respect of both supply and use considerations. Supply of the latter will, indeed, continue to expand until 2090, when its output will be some five times its level in 2000. The output of oil will, on the other hand, slowly decline from the 2050s, and its contribution to the annual hydrocarbon supply will ultimately fall to under 30%. Over the century as a whole, oil's share of the world's cumulative use of hydrocarbons will have been only 43%.

The Declining Importance Of Oil
The long-term future for oil will be squeezed – to a greater or lesser extent, depending on environmental and economic considerations – between the solidity of the annual 2.2 to 6.4 Gtoe coal supply over the whole of the 21st century and the powerful concurrent dynamics of the global gas industry. The future of oil will thus essentially be limited by demand considerations – including geo-political ones – so that potential resource (i.e. supply-side) limitations present only a secondary issue. This is shown in Chapter 3, which includes production curves covering the almost complete 200-year history of the industry from 1940 to 2140. Conventional oil is already well into its life cycle, but still has around 30 years to go to reach its peak. By contrast, non-conventional oil production has barely started. It has now begun to develop more rapidly, especially in Canada and Venezuela, but on an assumption that only 3,000 billion barrels of non-conventional reserves will be recoverable (from a resource base many times larger), then it will be 80–90 years before it reaches its potential peak production. This seems likely to be a little lower than that reached by conventional oil in 2030.

The two contrasting habitats of oil are, nevertheless, essentially complementary in respect of satisfying market demand, given that customers are interested only in the utility to them of the products they need and are wholly or largely indifferent as to how and from where the products they use have been derived. Non-conventional oil will, first, modestly supplement and, later, almost entirely substitute, so-called conventional production. The near-100-year period required for the full change from the one to the other can be portrayed as reflecting a slow, but continuing, process based on the importance of both economic geographical considerations and technological developments.

The oil component of energy production in the 21st century, however, is by no means a small or short-lived one. The supply increase shown for the first half of the century is at a rate made possible by already known reserves, reserves appreciations and new discoveries of conventional oil, plus the steadily rising flows of non-conventional oil. Nevertheless, the peak of oil production in the 2050s can be interpreted as being both retarded and lower than it might otherwise have been as a consequence of competition from other energy sources. The rate of decline in oil production after its peak in 2060 is, however, slow so that even by 2100 the oil industry is predicated to be larger than that in 2000. Nevertheless, by 2100, in the context of resource limitations and the strong competition for markets, it will then contribute rather less than coal to the global energy supply. By 2100 oil will, of course, already have been less important than natural gas for almost 60 years. Consequently, though oil's present geopolitical importance will undoubtedly continue in the first two decades of the 21st century, thereafter, as its contribution to the total energy supply progressively declines, so its political significance will rapidly fade.

Natural Gas – The Fuel Of The 21st Century

Natural gas will overtake coal as a global energy source early in the 21st century. This development reflects the tripling of proven gas reserves and the expansion of European and other markets for gas over the period since 1975. Gas thus entered the 21st century with a reserves-to-production ratio in excess of 60 years, with the implication that the expansion of production in the early decades will continue to be limited by demand, rather than by resources. Reserves in already discovered and producing fields could in themselves serve to keep global gas production growing at 3% or more a year until 2025, demand permitting, but additionally the continuation of additional discoveries is a certain prospect. This arises from the geographically broadening base of gas exploration activities and the continuing opportunities for the more intensive exploitation of existing gas-rich provinces, including some hitherto thought to be fully mature, including the Gulf of Mexico and the North Sea. The mid-point of the range of estimated additional reserves, indicates a volume about one-and-a-half times that of currently declared proven reserves. Proven reserves plus only half the estimated additional reserves are thus sufficient to support a conventional gas production curve which grows until at least 2050.

Chapter 4 shows how the 200-year curve of conventional gas supply from 1940 to 2140 will be complemented by an additional production curve for non-conventional natural gas. This latter curve emerges from a conservative assumption of ultimately recoverable reserves of 650Gtoe from coal-bed methane, tight formation gas, gas from fractured shales and ultra-deep gas. It does, however, exclude the possibility of any recovery from gas hydrates, the volumes of which have been estimated to be up to 20 times larger than all other non-conventional gas resources taken together. Large-scale non-conventional gas production is indicated to start in the 2020s and to become the more important component in the total gas supply before 2070. Its production will continue to grow to the end of the 21st century. In 2100 gas is predicated to supply about 50% of the world's carbon energy production and, over the 21st century as a whole, about 43% of the cumulative total.

Gas will undoubtedly be the fuel of the 21st century (as coal was of the 19th century and oil of the 20th), even given the limiting assumptions on the exploitation of gas resources specified above. These limitations, however, do inhibit the ability of the gas supply to grow sufficiently after 2060 to sustain a 2% a year expansion in overall non-renewable energy use. There is, however, a prospect by then for additional gas supplies from the initial exploitation of gas from hydrates. A further 60 years of continuing scientific advances and engineering capabilities would seem likely to provide enough time to achieve the recovery at least a small part of that massive potential source of energy, in the form of the world's preferred carbon fuel for both commercial and environmental reasons. It thus seems more likely than not to be brought to the market – assuming, of course, that the gas can then still compete with the alternative, renewable energy sources.

Oil and Gas as Renewable Resources
There is another possibility which suggests that the projections above for oil and gas may be less optimistic than is really the case. This arises from the issues of the origins of oil and gas. These fuels may be inorganic in origin.

A study on the long-term future of the world's energy supply which accepts without question the original 18th century hypothesis that oil and gas are generated from biological matter in the chemical and thermodynamic environments of the earth's crust must be considered to be somewhat fragile. There is, indeed, an alternative theory – already over 50 years old – which suggests an inorganic origin for oil and gas. This alternative view was widely accepted in the countries of the former Soviet Union where, it is claimed, large volumes of hydrocarbons are being produced from the pre-Cambrian crystalline basement. Recent applications of the inorganic theory, however, have also led to claims for the possibility of the Middle East fields being able to produce oil 'forever' and to the concept of *repleting* oil and gas fields in the Gulf of Mexico. More generally, it is argued by the proponents of the alternative theory that all giant fields are most logically explained by the inorganic theory, simply because calculations of potential hydrocarbon contents in sediments show that the organic materials present were too limited to generate the volumes of oil and gas that have been discovered or hypothesised.

The high significance of the alternative theory of the origin of oil and gas to the issue of energy supplies in the 21st century is self-evident. Instead of having to assume that reserves of oil and gas are already accumulated in a finite number of fields in so-called oil and gas plays within defined petroliferous provinces, the possibility emerges of evaluating the presence of hydrocarbons over geographically much larger areas of the earth's crust. In this context they become essentially renewable resources no matter what demand developments may emerge. If fields do replete because the oil and gas extracted from them is abyssal and abiotic (based on chemical reactions under specific thermodynamic conditions deep in the earth's mantle), then extraction costs should not rise as production from such fields continues for an indefinite period. Nor would estimates of reserves, reserves-to-production ratios and annual rates of discovery and additions to reserves justify the importance attributed to them in evaluating future supply prospects under the organic theory of oil and gas derivation. In essence, the 'ball park' hitherto used for considering the issues relating to future supplies of oil and gas would no longer remain appropriate. Geographically more extensive and geologically deeper habitats can be brought under consideration in what is, in effect, a much bigger 'ball park'.

References

Anderson, D. (2001), "Energy and economic prosperity." *World Energy Assessment*, New York, UNDP.

Anthrop, D.E. (2003), "Hydrogen and Fuel Cells," *Oil and Gas Journal*, Vol.101.39, pp.10–15.

Bartsch, U. and Müller, B. (2000), *Fossil Fuels in a Changing Climate*, Oxford, O.U.P.

Bossel, H. (1998), *Earth at a Crossroads; Paths to a Sustainable Future*, Cambridge, C.U.P.

Bradley, R.L. (2003), *Climate Alarmism Reconsidered*, London, Institute of Economic Affairs.

Campbell, C.J. (2003), *The Essence of Oil and Gas Depletion*, Brentwood, Multi-Science Publishing.

Claes, D.H. (2001), *The Politics of Oil Producer Cooperation*, Boulder, Westview Press.

European Commission (2003), *Towards a Hydrogen-based Energy Economy*, Brussels, EU.

European Environment Agency (2002), *Energy and Environment in the European Union*, Brussels

Freund, P. (2002), "Technology for avoiding CO_2 emissions," *Proceeding of the 17th World Petroleum Congress*, Rio de Janeiro, Vol.5, pp.11–21.

Gerholm, T.R. Ed. (1999), *Climate Policy after Kyoto*, Brentwood, Multi-Science Publishing.

Greenpeace (1997), *Putting a Lid on Fossil Fuels*, London, Greenpeace.

Griffiths, J. (2001), "Fuel cells – the way ahead," *World Petroleum Congress Report*, London, ISC Ltd.

Grimston, M.C. and Beck, P. (2002), *Double or Quits: the Global Future of Nuclear Power*, London, R.I.I.A.

Heinberg, R. (2003), *The Party's Over; Oil, War and the Fate of Industrialized Societies*, British Columbia, New Society Publishing.

Hoffman, P. (2001), *Tomorrow's Energy; Hydrogen, Fuel Cells and the prospect for a Cleaner Planet*, Cambridge, Mass., Cambridge University Press.

I.E.A. (1997), *Energy Technologies for the 21st Century*, Paris, OECD.

I.E.A. (2002a), *Flexibility in Natural Gas Supply and Demand*, Paris, OECD.

I.E.A. (2003a), *South American Gas: Daring to Tap the Bounty*, Paris, OECD.

I.E.A. (2003b), *World Energy Investment Outlook*, Paris, OECD/IEA.

I.P.C.C. (2001), *Climate Change*, Cambridge, Cambridge University Press.

Jean-Baptiste, P. and Ducroux, R. (2003), "Energy policy and climate change," *Energy Policy*, Vol.31.2. pp.155–166.

Klare, M. (2002), *Resource Wars: the New Landscape of Global Conflict*, New York, Metropolitan Press.

Kleveman, L. (2003), *The New Great Game: Blood and Oil in Central Asia*, London, Atlantic Books.

Maritis, G. (2003), "CO_2 sequestration adds new dimensions to oil and gas production," *Oil and Gas Journal*, Vol.101.9. pp.39–44.

Meadows, D.H. et al (2002), *Beyond the Limits: Confronting Global Collapse*, Washington DC, Chelsea Green.

Mitchell, J. et al (2001), *The New Economy of Oil*, London, R.I.I.A.

Odell, P.R. (1984), "The oil and gas resources of the Third World importing countries and their exploration potential," *U.N. Development Research and Policy Analysis Division*, New York.

Odell, P. R. (1990), 'Continuing long-term hydrocarbons dominance of world energy markets: an economic and societal necessity,' *Proceedings of the World Renewable Energy Congress*, Oxford, Pergamon Press.

Pachin, J.C. (2003), "Coal has CO_2 capture opportunities," *AAPG Explorer*, Vol.24.8, pp.36–8.

Randall, S.J. (2002), "Energy Security in the 21st Century." *Geopolitics of Energy*, Vol.24.4, April.

Rifkin, J. (2002), *The Hydrogen Economy*, London, Penguin Books.

Thomas, R. and Ramberg, B. (Eds.) (1990), *Energy and Security in the Industrializing World*, Lexington, U.P. of Kentucky.

Torp, T.A. (2001), "Carbon sequestration: a case study," *17th World Petroleum Congress Report*, London, ISC Ltd.

Williams, R.H. (1998), "A technological strategy for making fossil fuels environment and climate friendly," *World Energy Council Review*, September, pp.59–67.

World Petroleum Congress (2002), "Natural gas: clean energy for half-a-century," *Proceedings*, Vol.4. Forum 14.

Chapter 1: The Global Energy Economy in 2000: and its Prospects for the 21st Century

The Historical Background

In a world which remains overwhelmingly dependent on the use of carbon fuels for its energy needs, the issue of their potential availability must inevitably be of concern, even though recurring fears of impending scarcity in previous decades have all proved groundless. In this latter respect, the early 1970s pessimism of the Club of Rome in its study, *The Limits to Growth*, (Meadows, 1972) presented the dangers for a world which, it alleged, faced relatively near future physical constraints on the production of adequate supplies of energy. This was in the context of what was then expected to be the continuation of the post-World War II high rate of growth of an average of 5% or more a year in the demand for coal, gas and, especially, oil. The perceived danger was, moreover, compounded by the seeming absence at that time of any alternative low cost sources of energy which could replace the use of carbon fuels within any relevant time-span.

Some 30 years later, the validity of both of these central issues has been eliminated. The rate of increase in the global use of energy has since 1973 declined to below its very long-term 1860–1945 trend of 2.2% a year (see Figure 1.1), and thus to well below the growth rate of nearly 5% a year over the 28 years after the end of the second world war. This high rate of growth in energy demand from 1945 to1973 can, however, be shown to be the result of a temporally unique set of circumstances, viz. that the economies of almost all of the world's nations were then going through heavily energy-intensive periods of development in the context of a close relationship between the rates of increase in energy use and the pace of economic development – as shown in Figure 1.2, (Odell, 1973).

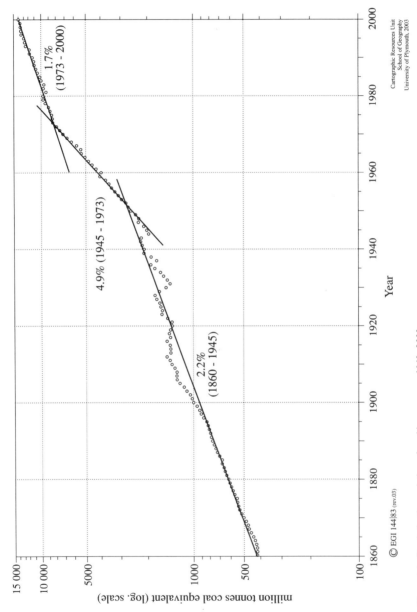

Figure 1.1 Trends in the evolution of world energy use, 1860–2000

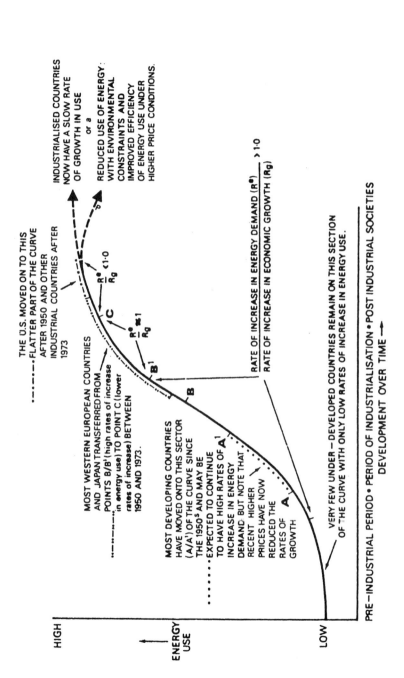

Figure 1.2 Relationship between Energy Use and Economic Development over Time

The world's growing use of energy depends on the changing relationships over time between energy use and economic development in different economies. Note that the world's industrialised nations (the USA, Western Europe etc.) have already moved off the steepest part of the curve whilst most developing countries are progressing along it. Very few countries are moving from low Re/Rg ratios to higher ones. Post-industrial societies which pay increasing attention to environmental considerations and to questions of efficiency in energy use could expand economically whilst using less energy overall.

First, the small number of rich countries in the West were in the later stages of the traditional industrialisation process, marked by an emphasis on products with high energy inputs, such as motor vehicles, durable household goods and petrochemical products (Schurr and Netschert, 1977). As a result, the use of energy on the production side of these economies greatly increased. Simultaneously, the increase in the per capita use of energy on the demand side of these countries' economies was even more dramatic. This was the consequence of a number of inter-related factors, viz. the suburbanisation of cities and the increasing length of the journey to work, the switch from public to private transport involving the mass use of motor cars, the expanded availability of leisure time accompanied by the 'annihilation of space' in the public's use of such time, the mechanisation of households by the use of electrically powered equipment and the achievement of much higher standards of comfort (by heating and cooling) in homes and other buildings (Leach et al, 1979). Cheap energy was one of the main bases for the 'revolution of rising expectations' on the part of the populations of the rich countries. The progress of that revolution in the 1950 and 1960s was thus marked by a rapid increase in energy use in the industrialised world (Odell, 1974).

Second, many of the same factors positively influenced the rate of growth in energy use in the centrally planned economies of the former Soviet Union (FSU) and eastern Europe, despite the differences of ideology and of economic and political organisation between East and West (Park, 1979). This was particularly the case in respect of the industrialisation process in which all these countries participated as a matter of deliberate policy – and even with a special emphasis on the rapid expansion of heavy energy-intensive industry. To a smaller but, nevertheless, still significant extent, consumers in the centrally planned economies also increased their levels of energy use in this period as a result of higher real living standards and changes in lifestyle. Electrification, a particularly energy-intensive process, was a declared central aim of such planned economies – not least because Lenin had specified it as a necessary part of the evolution to communism (Lenin, 1966). Thus, all these countries were at this time also moving up the steepest part of the curve representing the relationship between energy use and economic development (Dienes and Shabad, 1979).

Meanwhile, most of the countries of the developing world had been moving off the lowest part of that curve in the 1950s and 1960s, as these (for the large part) newly independent nations also pursued policies of deliberate industrialisation. This process was, indeed, viewed as the panacea for, and the *sine qua non* of, economic progress. The types of industry which were early to be established were either energy-intensive heavy industries – such as iron and steel, metal fabrication, vehicle production and cement – or industries such as textiles and household goods which were also relatively energy intensive. Such industrialisation policies were, moreover, necessarily accompanied by the rapid urbanisation of the population with its much enhanced energy requirements. In the process, peasants and landless agricultural labourers were transferred from their low-energy ways of rural living (in which most of the energy required was collected rather than purchased), to lifestyles in the city or urban environment which, no matter how poor the living standards achieved turned out to be, were much more demanding in their requirement for energy, generally, and in their use of electricity and petroleum products, in particular (Dunkerley, 1981).

It was essentially the temporal coincidence of these fundamental societal developments across a very wide range of countries in the world in the 1950s and 1960s which caused the abnormally high rate of growth in global energy use over that period. In passing, however, it is worth noting that the US was already moving off the steepest part of the curve in the late 1960s and early 1970s as its rate of growth in energy use started to fall away under the impact of structural changes in its economy, viz. when the service sector started to grow more quickly than manufacturing (Schurr and Netschert, 1977). Because of the relative importance at that time of the United States' use of energy – at almost 35% of the world total – this development exercised a downward pull on the global rate of increase in energy use.

That phenomenon, though later to become of great significance in defining energy use growth rates, remained unobserved at the time as its impact was temporarily masked by another powerful factor at work influencing the rate of increase in energy use in the two decades prior to the price shock in 1973/4. This was the continuing decline in the real price of energy in the 20 years since 1950 (Adelman, 1972) when the (real) price of Saudi Arabian light crude oil fell by over 60% from $4.25 to $1.60 (as shown in Figure 1.3). This reflected the impact of the technologically efficient and politically powerful oil companies which, at that time, were not only able largely to ignore the interests of the oil-producing countries in the exploitation of their resources, but were also successful in persuading the importing countries of the assured continuity of these low cost oil supplies (Odell, 1971).

This decline in the oil price necessarily brought about a falling market price for all other sources of energy, although not to the same degree as the fall in the price of crude oil itself, when compared with the reductions in price to the final consumer of energy. The fall in energy prices was especially important in the western industrial countries with their open – or relatively open – economies under their post-war liberal trading regimes. In these countries, the production of their indigenous sources of energy – namely, coal in Western Europe and Japan, and oil and natural gas in the US – either had to be reduced in price to enable them to compete with cheaper imported oil, or the industries concerned had to be cut back or eliminated (Manners, 1971; Odell, 1971 and 1975).

Thus, both the real decline in the cost of energy over this period and the perception among energy users that energy was cheap and getting cheaper – so that it appeared hardly worth worrying about in terms of the care and efficiency with which it was used – created conditions in which the careless and wasteful consumption of energy became a hallmark of both the technological and behavioural aspects

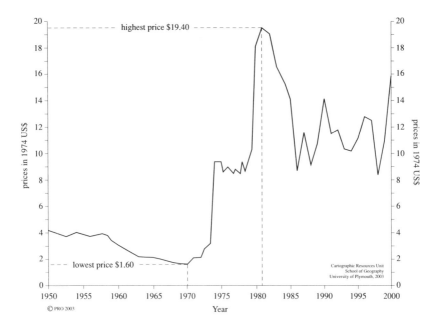

Figure 1.3 The Price of Saudi Arabian/Dubai Light Crude Oil, 1950–2000

of societal and economic developments. There was a consequent emphasis on systems of production, transport and consumption which were quite unnecessarily energy intensive (Darmstadter, 1977). Given these conditions, then a reaction of significant proportions could be expected once the real price curve started to move up. This phenomenon, as shown in Figure 1.3, dated from 1970 when oil exporting countries required the companies to increase prices in order to generate more revenues for their governments. Just a few years later, however, in 1973–74, it took a massive upward leap with the so-called 'first oil price shock' arising out of political developments in the Middle East (Odell, 1986). By the time of the second price shock in 1979–1981, also related to political developments (Odell, ibid.), structural changes in the energy-careless and wasteful systems were already well under way. They have since both intensified and become geographically much more diffuse (Schipper and Meyers, 1992).

In other words, the order of magnitude jump in oil and the lower, but still significant, increases in other energy prices in the 1970s terminated the perception of energy as a near-costless input to economic and societal developments in most parts of the world (Odell and Valenilla, 1978). Consequently, as shown in Figure 1.1, the rate of increase in energy use has since 1973 been brought back to well below its historical long-term level; to only 1.7% per year. The most important element in the change has been the quite dramatic decline in the energy intensity of economic activities in the western industrialised countries (Schipper and Meyers, 1992). Here the motivation to achieve energy savings was greater than elsewhere, while the ability to do so has been higher (Fritsch, 1982) as both behavioural and technological components contributed to this development.

Behaviourally, users responded to the higher prices by taking steps which saved energy (such as turning down room thermostats and using motor vehicles more effectively). At the same time they were subjected to energy efficiency and conservation campaigns. It is, however, the technological component which has been much more effective in reducing energy use growth rates (Brookes, L., 2000). This is largely because there was so much excess to work out of the systems created in the energy-careless and thoughtless technology of the pre-1970 period. More efficient energy processes in factories, more efficient lighting in offices, better insulation in buildings, the development of motor vehicles, planes and ships which gave more kilometres per litre of fuel and the expansion of systems of electricity production which are inherently more energy efficient (such as natural gas fuelled combined-cycle generation), combined to produce significantly lower rates of energy use per unit of output of goods and services (Schipper and Meyers, 1992). In Western Europe, for example, the energy intensity of economic growth from 1973 to 1980 was less than half that of the preceding 10 years (International Energy Agency, 1984). Since 1980 the ratio continued to fall, albeit at a slower rate, especially since energy prices moved down sharply in real terms from 1986 to 1997 (see Figure 1.3).

The diffusion of more efficient energy systems and infrastructure to the modern sectors of most developing countries has necessarily been slower because of the relative scarcity in such countries of the inputs that are required to implement the changes – notably investment capital and expertise (Smil and Knowland, 1980). Some diffusion, however, took place under the powerful stimulus of the high foreign exchange costs of energy imports, particularly oil, on which developing countries had come to depend to an even greater degree than most industrialised countries (Dunkerley, 1981; Pachauri, 1988). In such cases the alternative to more efficient energy use in the short-term became the affordability of sufficient energy to keep systems going, as limits had to be placed on the amount of oil that could be imported. Nevertheless, enhanced levels of energy-utilisation efficiency were eventually achieved in the developing world, though to nothing like the same degree as in the industrialised countries (Anderson, 2001).

This contrast is an important consideration in relation to the longer-term evolution of global energy demand, given that the percentage of world energy used in these countries will continue to increase steadily. Indeed, it has already increased from 12.5% in 1980 to its present 34.5% under the joint impact of their high rates of population growth and of ongoing changes in the structure of their economies. These, in the main, are still going through the most energy-intensive period of development (Anderson, *ibid.*, Desai et al, 1987). The countries of the developed North, on the other hand, though now using more energy per capita than in 1973, have already experienced slow growth rates in energy use for over 25 years (Hoffman and Johnson, 1981; Guilmot, 1986). In a few cases (for example, Denmark, Germany and the UK) they have, over the last few years, even achieved small reductions in their per capita use of energy.

The countries which until the mid-1990s had centrally planned economies (viz. Russia, the other former Soviet republics and the countries of Eastern Europe plus China, Vietnam, Cuba and a few other Third World countries) lie between the industrialised and the developing nations in respect of their historic energy use patterns and the prospects for change (Dienes and Shabad, 1979; Park, 1979; Hoffman, 1985). In general, they had, until the demise of their Communist regimes and/or ideology, done significantly less well than the market economies in saving energy – partly for economic (pricing) reasons, and partly for technical and organisational reasons. Since the regime changes (as in the former Soviet Union and Eastern Europe) or in the context of a significant relaxation of centralised planning (as in China and Vietnam), the importance of energy efficiency and conservation have become much more generally recognised as requiring a much higher priority in national energy policies (Odell, 2001; Considine and Kerr, 2002).

Prospects for Energy Demand

Overall, structural changes in attitudes and policies towards the energy sectors of the economies of countries in all parts of the world, had by 2000 generated long-term prospects for a continuing rate of increase in energy demand as low as that in the last quarter of the 20th century – even in the context of renewed higher rates of global economic growth. By 2000 huge differences had emerged between the early 1970s' expectations of future energy demand (Odell and Rosing, 1980/3) and that which was actually generated by 2000. The continuation post-1973 of the abnormally high growth rate in energy use of the years since the late 1940s would have required an annual global supply of energy of over 22 gigatons of oil equivalent (Gtoe) by 2000. In sharp contrast, as shown in Figure 1.4, the actual growth rate of only 1.7% a year from the 1973 base led to a use of energy in 2000 of less than 10Gtoe.

As already shown in the arguments above, the probability of a future reversion to the temporary 25-year high rate of growth between 1945 and 1973 is very low. Indeed, the same combination of conditions which applied in that period *cannot* occur again. Instead, consideration of future energy supply requirements can now, with a high degree of confidence, be based on a much longer-term global demand expansion which does not exceed an average of 2% a year, even without taking into account the impact on demand of the increasing attention now being paid to environmental concerns, especially with respect to global warming and its perceived links with CO_2 emissions from carbon based energy sources.

Moreover, energy alternatives to the use of carbon fuels have grown modestly in the past 30 years. Thus, from providing around 0.65Gtoe (in heat value equivalent) in 1971, they accounted for over 0.9Gtoe in 2000, compared with the almost 50% decline over this period in the use of traditional, generally non-commercial, energy sources. Over the period to 2020 a somewhat faster rate of growth in the availability of alternative energies – of about 2.5% a year – should be achievable, as a consequence of the rapidly evolving technologies of direct and indirect solar energy based production. The availability by then of more than 2.1Gtoe of alternative energies will thus modestly restrain the expansion of demand for carbon fuels. The contribution of carbon fuels to total energy supplied will fall to less than 85%, from a 90% share in 2000. Beyond 2020 this relative shift to the use of alternative energy sources will slowly but steadily intensify, increasing their contribution to total energy supply to over 25% by 2070 and to 34% by 2080. Thereafter, modest declines in the supply of carbon energy, will enable renewables to increase their contribution to 39% by 2090 and to 43% by 2100. The dynamics of the process are shown in Figure 1.4.

Thus, an examination of the long-term future availabilities of carbon fuels can be orientated to a 'highest' case requirement for a significantly less than 2% per annum growth rate. The prospects for the future reserves and resources of carbon fuels that now have to be examined are, as a consequence, fundamentally different from the prospects that were viewed in such a sombre way in the 1970s (Odell and Rosing, 1980/3).

Indeed, the actual developments over the past 30 years in the global energy economy make the forecasts which were made in the early 1970s seem quite extraordinary.

For example, the 1972 forecast by H.R. Warman, then Chief Geologist of British Petroleum, was for a cumulative use of oil over the last three decades of the 20th century of almost 1,750 billion barrels (240Gtoe), with the use of oil in the year 2000 indicated by him at almost 140 billion barrels (19Gtoe) (Warman, 1972). Such forecasts by oil industry spokesmen were then compared with the cumulative production of oil over the whole of the previous history of the industry of 'only' 400 billion barrels (55Gtoe) and an output in 1970 of 16.6 billion barrels (2.3Gtoe). In this context the world's oil needs within the relatively short 30-year period to 2000 seemed forbidding and formidable. The views thus led to a high degree of pessimism on the ability of the industry to find not only the 1,750 billion barrels which, it was anticipated, would be used by 2000, but also the minimum

The Global Energy Economy in 2000: and its Prospects for the 21st Century

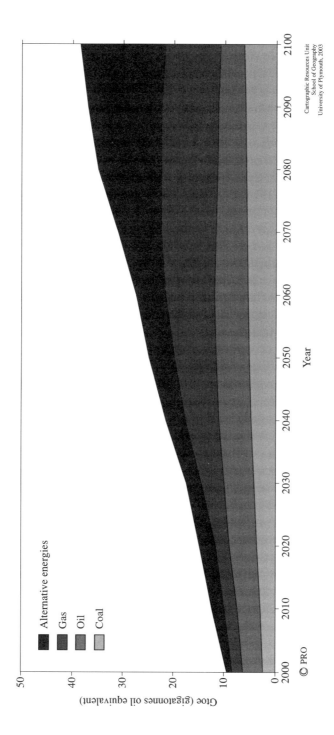

Figure 1.4 Trends in the Evolution of Energy Supplies by Source and per Decade in the 21st Century

2,000 billion additional barrels of oil which would have to be found by the end of the century in order to provide a 20-year reserves-to-production ratio for the forecast level of production over the next 30 years to 2030 (Odell and Rosing, 1980/3).

It is worth emphasising that this pessimistic view of the future of oil was not just 'simply wrong', but that it also generated national and international energy policies which produced not only much higher energy prices, but also needlessly large investments in high cost alternatives to oil, coal and natural gas – especially for the development of nuclear power (Beck, 1994)

Reality, of course, has turned out to be massively different from the expectations of Warman and others, such as A. Hols, the Chief Geologist of the Royal Dutch/Shell Group (Hols, 1972). Global energy use by 2000, indeed, proved even much more parsimonious than the more modest alternative prospects which the present author had forecast in the early 1970s (Odell, 1973), viz. a predicted 'need' for oil reserves additions by 2000 of 1,000–1,250 billion barrels (137–171Gtoe) and for an industry which would need to produce about 40 billion barrels (5.5Gtoe) a year by the end of the century. In the event, cumulative oil production from 1971 to 2000 totalled just 680 billion barrels (92Gtoe). Production in the year 2000 was less than 26.5 billion barrels (3.8Gtoe), only modestly higher than the 17 billion barrels (2.4Gtoe) produced in 1971.

In other words, the production base-line at the end of the 20th century from which to look at the future requirement for oil, was less than 16% of the conventional forecasts of the 1970s. Concurrently, the demand for gas doubled, though from a much lower starting point in 1971 of only about 1.1Gtoe, to almost 2.2Gtoe in 2000. Coal, however, which was relatively much more important than gas in 1971 (at 1.8Gtoe), suffered from competition from both oil and gas, so that its use grew by only 20% by 2000 to 2.2Gtoe. Taking the three carbon fuels together, growth in use over 30 years was just over 55%. In the light of these developments it is thus hardly surprising that expectations for future requirements for carbon fuel resources – and for the supply potential associated with them – can now be so much more relaxed than they were 30 years ago.

Carbon Fuel Resources

This subject of the book is limited, in by far the greater part, to the so-called non-renewable sources of energy. It thus excludes detailed consideration of alternative energy sources and, in particular, of issues relating to their long-term availability. These limitations imply some conceptual problems – as already suggested in the first part of this chapter. Most specifically, they beg the question of the nature of resources, given that 'resources are defined by man, not by nature' (Rees, 1985). The concept 'resource' thus presupposes that a 'planning agent' is appraising the usefulness of his environment for the purpose of obtaining a certain end (Ciriacy-Wantrup, 1952). This, in turn, implies that before any source can be classified as a resource, two basic conditions must be satisfied, viz.

- first, that knowledge and skills must exist to allow its extraction and use;
- and second, that there must be a demand for the materials or services produced.

If either, let alone both of these conditions remains unsatisfied, then the physical substances or occurrences at or within the earth's surface remain 'neutral stuff' (Zimmerman, 1951). In other words, resources involve human ability and need. The mere physical presence of a "resource" has zero significance; as, for example, with most of the remaining coal underlying large areas of Europe for which there is no demand, or with deep ocean gas hydrates which currently cannot be recovered.

However, because human abilities and needs are highly dynamic, today's energy resources incorporate some of yesterday's 'neutral stuff'. This is the case, for example, of oil in reservoirs lying under continental shelves and slopes where exploitation was, until the last third of the 20th century, virtually impossible. They also, however, exclude some of yesterday's valuable reserves; for example, coal in areas where demand which was previously expanding, is now declining or even extinct, and in areas where exploitation can no longer be undertaken as a result of concerns for the local environment. Thus, while tomorrow's energy resources of carbon fuels will *undoubtedly* include components relating not only to already known, but as yet unproducible, reserves (e.g. deep oil and gas), they will also incorporate reserves from today's unknown potential in still unexplored habitats.

In order to incorporate both human ability and human need into an evaluation of carbon fuels prospects, two basic assumptions must be made, viz.

- first, that neither additions to knowledge of the world's geology,

nor the development of technology with respect to the exploitation of coal, oil and gas have come to an end. We take these to be self-evident truths (USGS, 2000; World Petroleum Assessment, Reston, Government Printing Office).
- second, that the world's growing population and economies will generate a continuing increase in the demand for energy over the long term (United Nations, 2001a).

Even within the framework of an exponential value overall for this latter parameter of under 2% a year until 2050, the annual average growth rate in the demand for coal can be assumed to be below this figure for environmental reasons and thus below the rate at which the world's coal resources could otherwise be economically exploited. On the other hand, the growth rate in the total demand for oil may be assumed as able to grow exponentially at 1.5% a year for only as long as the evolution of reserves availability makes this possible, viz. to about 2060. For gas, both resources availability and environmental advantages would sustain a close to 3% per annum growth rate over the first half of the century, but this will be restrained by demand limitations. Post-2050, as the world's population stabilises at ±9 billion inhabitants, the required growth rate for energy seems likely to fall gradually, decade by decade, to a low of no more than 1.2% per annum by the last decade of the century. Should demand pressures on remaining hydrocarbons resources, in general, and on oil resources, in particular, be ameliorated, then the time period over which their production levels could be increased will be extended. Meanwhile, coal seems likely to remain constrained by environmental concerns for most of the century.

Arising out of this approach to the future supply of coal, oil and natural gas, the expansion of the use of renewable reserves is, in essence, treated as the balancing item to ensure that the emerging growth rates in energy use as set out above can be achieved. Given the continuation of significant cost differentials in favour of exploiting carbon fuels over renewables, this seems to be a reasonable proposition for at least the first quarter of the 21st century. In the second quarter of the century, the more rapid expansion of the demand for renewable sources of energy – under decreasing cost conditions arising from technological advances and economies of scale – may so enhance competition for carbon fuels that the growth rate in the use of coal and oil could well be restrained by demand-side competition, rather than by environmental, concerns. It could even lead to the annual growth rate for oil plus gas supply falling below the 2% level which, as will be

The Global Energy Economy in 2000: and its Prospects for the 21st Century

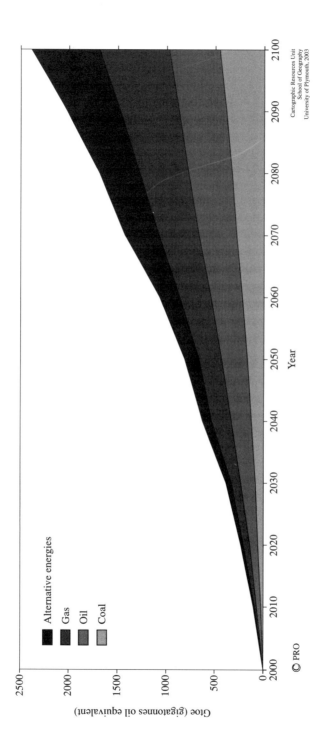

Figure 1.5 Cumulative Supplies of Energy, by source, in the 21st Century

shown below (see Chapters 3 and 4), remains technically feasible. This development, however, seems more likely after 2050. From the mid-century, there will be a steadily increasing degree of probability that alternative energies will substitute for carbon fuels on economic grounds. Under such circumstances, the eventual decline curves in the supply of oil (post-2060: see Fig. 3.5) and gas (post-2090: see Fig. 4.2) could, instead of being a function of reserves availability, well emerge as the result of demand limitations.

The end result of taking these variables into account is shown above in Figure 1.4. The calculations on which this graph is presented are set out Chapters 2, 3 and 4. Here, attention is simply drawn to the main outcome of the analysis, viz. that carbon fuels collectively remain more important than all alternative sources put together until after 2100, but noting, nevertheless, that almost all the growth in energy supplies after 2070 will come from alternative sources. The cumulative data on 21st century energy use are presented in Figure 1.5. This shows that alternative energies will still have provided only 18% of total energy used in the first half of the 21st century and 31% by 2100. Carbon fuels will thus dominate this century's energy supply; albeit with radically different contributions by coal, oil and gas compared with what they were in the 20th century.

References

Adelman, M.A. (1972), *The World Petroleum Market*, Baltimore, Johns Hopkins University Press.

Anderson, D. (2001), "Energy and economic Prosperity," *World Energy Assessment*, New York. U.N.D.P.

Beck, P. (1994), *Prospects and Strategies for Nuclear Power*, London, R.I.I.A.

Brookes, L. (2000), "Energy Efficiency Fallacies Revisited", *Energy Policy*, Vol.28.4, pp349–60.

Ciriacy-Wantrup, S.V. (1952), *Resource Conservation, Economics and Policies*, Berkeley, University of California Press.

Considine J.I. and Kerr, W.A. (2002), *The Russian Oil Economy*, Cheltenham, Elgar Publishing.

Darmstadter, T. et al (1977), *How Industrial Societies use Energy*, Baltimore, Johns Hopkins University Press.

Desai, H. et al. (Eds.) (1987), "Special Lesser Developing Countries Issue", *Energy Journal*, Vol.8.

Dienes, L. and Shabad, T. (1979), *The Soviet Energy System*, Washington D.C., Winston Press.

Dunkerley, J. (1981), *Energy Strategies for Developing Nations*, Baltimore, Johns Hopkins University Press.

Fritsch, B. (1982), *The Energy Demand of Industrialised and Developing Countries*, Zurich, Institute of Technology.

Guilmot, J.F. (1986), *Energy 2000*, Cambridge, Cambridge University Press.

Hoffman, G.W. (1985), *The European Energy Challenge; East and West*, Durham N.C., Duke University Press.

Hoffman, J. and Johnson, B. (1981), *The World Energy Triangle*, Cambridge, Ballinger Press.

Hols, A. (1972), "Future energy supplies to the Free World," *E.I.U. International Oil Symposium*, London, pp.1–24.

I.E.A. (1984), *World Energy Outlook*, Paris, OECD.

Leach, G. et al (1979), *A Low Energy Strategy for the United Kingdom*, London, I.I.E.D.

Lenin, V. I. (1966). *Collected Works of V. I. Lenin*, Vol.31, Moscow, Progress Publishers.

Manners, G. (1971), *The Geography of Energy*, London, Hutchinson.

Meadows, D.H. et al (1972), *Limits to Growth*, Washington, Potomas Association.

Odell, P.R. (1971), *Oil and World Power*, London, Penguin Books.

Odell, P.R. (1973), "The future of oil: a rejoinder." *Geographical Journal*, Vol.139.3, pp.436–454.

Odell, P.R. (1974), *Energy: needs and resources*, Basingstoke, MacMillan Education.

Odell, P.R. (1975), *The Western European Energy Economy*, London, Athlone Press.

Odell, P.R. (1986), *Oil and World Power*, London, Penguin Books.

Odell, P.R. (2001), *Oil and Gas: Crises and Controversies, 1961–2000*, Vol.1, *Global Issues*, Brentwood, Multi-Science Publishing. pp.617–24.

Odell P.R. and Rosing, K.E. (1980/3), *The Future of Oil, 1980–2080*, London, Kogan Page.

Odell P.R. and Vallenilla, L. (1978), *The Pressures of Oil*, London, Harper and Row.

Pachauri, R.K. (1988), "Energy and growth: beyond the myths and myopia," *The Energy Journal*, Vol.10.1, pp.1–20.

Park, D. (1979), *Oil and Gas in the COMECON Countries*, London, Kogan Page.

Rees, J. (1985), *National Resources: Allocation, Economics and Policy*, London, Methuen.

Schipper, L. and Meyers, S. (1992), *Energy Efficiency and Human Activity*, Cambridge, Cambridge University Press.

Schurr, S. and Netschert, B. (1977), *Energy in the American Economy, 1850–1975*, Baltimore, Johns Hopkins University Press.

Smil, V. and Knowland, W.E. (1980), *Energy in the Developing World: the Real Energy Crisis*, New York.

United Nations (2001a), *Sustainable Energy, Shifting towards a Development Path*, New York, UN Series no.38.

United States Geological Survey (2000), *World Petroleum Assessment*, Reston, Government Printing Office.

Warman, H.R. (1972), "The future of oil," *Geographical Journal*, Vol.138.3, pp.287–97.

World Petroleum Congress (2002), "New hydrocarbons provinces of the 21st century," *Proceedings*, Vol.2, pp.87–176.

Zimmerman, E.W. (1951), *World Resources and Industries*, New York, Harper and Row.

Chapter 2: Global Coal Resources and Production Prospects

Long-term Availability

There is little controversy concerning the world's large and widely distributed coal resources. The earliest attempt, in 1913, to survey and report on global resources for consideration at the 12th International Geological Congress in Toronto lacked adequate geographical cover and standardised criteria for measurements. Nevertheless, in spite of these limitations, it still provided the main basis for the subsequent effort in 1936, for the 1st World Power Conference in London, to systematise and update the evaluation of the world's coal wealth. This evaluation indicated a total of 8,166 x 10^9 tons of likely recoverable resources – excluding a claim for additional resources in China of roughly the same magnitude. The regional breakdown of the data is shown in Table 2.1. In this the domination of the coal resources of North America and Europe (including Russia) stands out clearly. The data of that time reflected, of course, the continuing importance of these two continents in the global coal exploration and exploitation process: with this, in turn, reflecting the concentration of coal use (and hence of interest in coal reserves) in the two regions.

Table 2.1 also shows data as closely comparable as possible from the mid-1970s and the mid-1990s. In the 1970s the world's coal resources which were designated to be exploitable were declared at over 11,000 x 10^9 tons. The 35% increase compared with 1936 was almost entirely the result of a massive uprating of the resources of the Soviet Union. By contrast, the resources of North America had been downgraded in the intervening 40 years as a consequence of the shift in the country's energy economy away from coal to oil and natural gas (Manners, 1971; Schurr and Netschert, 1977).

Table 2.1: Assessments of world coal resources, 1936–1996

tons of coal equivalent x 10^9

Region	1936[1]	1976[2]	1996[3]
North America	4,149	3,322	883
Europe (including the FSU)	2,441	5,481	3,302
Asia	1,168	1,678	1,771
Africa	231	221	153
Australasia	173	686	89
Latin America	2	35	47
Total	8,166	11,423	6,246

Source:
1 Zimmermann, E. W., *World Resources and Industries*, Harper and Row, New York, 1951
2 Grossling, B. F., *World Coal Resources*, Financial Times Business Information Ltd, London, 1979
3 Rogner, H-H., *An Assessment of World Hydrocarbons Resources*, IIASA, WP96/56, 1996

By the mid-1990s, estimates of the world's coal wealth had dropped to a volume which was about 25% below that of 1936, even though coal production and use over the intervening 60 years was no more than 250 x 10^9 tons (equal to about 3% of the estimated coal resources in 1936). The re-calculated 'resource base' was 6,246 x 10^9 tons. Of this lower total in 1996, however, almost 50% was defined as unlikely to be exploitable except *in extremis*. North America's resources, in particular, were reduced by two-thirds, Europe's (including those of the FSU) by 40%; and Australasia's by over 85%. Even so, the effective "resource base" (of 3000 x 10^9 tons) constituted some 850 years of coal supply at the 1996 level of global production.

Nevertheless, as early as the 1970s fears were expressed about the future availability of recoverable coal, given the expectation of a significant and continuing expansion of demand for it, in the context of the widely held belief at that time in the scarcity and the anticipated increasingly high price of oil. At the 1976 1st International Coal Exploration Symposium the chairman spoke of 'an emerging and revitalised world role for coal... based on the enormous world reserves of coal when compared to gas and crude oil.' (Muir, 1976) In these circumstances, the 50% increase in coal use over the 25 years from 1950 was contrasted with an expected close-to-100% increase in use between 1975 and 2000, by which date a potential annual use of coal by the turn of the century would be almost 5 x 10^9 tons (3.3Gtoe). Coal and nuclear were presented as the only two major energy sources which at the end of the century would be able to fill the 'energy gap' created by the run-down in world oil and gas resources (Green and

Gallagher, 1980). Other contributors put the position even more dramatically; viz. "the role of coal has been undeservedly belittled for a short period recently, but it is now expected to rescue mankind during an energy crisis" (Muir, 1976).

In that context, the world's proven reserves of coal – at just under 700×10^9 tons (465Gtoe) in 1976 (see Table 2.2) were shown to have emerged from an overall degree of exploration of only about 7.5% of the resource base and there was a call from Prof A. K. Matveev of Moscow State University for "a systematic and intensified effort by an authoritative international commission to secure a worldwide evaluation of the total geological reserves of coal by types and qualities, within the framework of agreed parameters relating to depth and critical thickness of seams." In this way, it was argued, the problem of the future availability of coal, following the depletion of "presently defined large amounts of the total known coal reserves by the beginning of the 21st century," could be resolved (Muir, *ibid*).

Although this view was a misrepresentation of reality – in that the expected doubling of annual coal use from 2.5×10^9 tons (1.7Gtoe) in 1975 to 5×10^9 tons (3.3Gtoe) in 2000 would use up less than 100×10^9 tons (66Gtoe) of the estimated reserves in 1976 of 690×10^9 tons – it illustrated the attitude towards the fear of scarcity which was prevalent at that time. This non-rational view of the future availability of energy resources was presented not only by the leaders of the industries concerned, but also by reputable scientists and professionals.

Table 2.2: Proven coal reserves, 1936–2001

Region	1936[1]	1976[2]	1996[3]	2001[4]
		tons x 10^9 of hard coal equivalent		
North America	421	200	181	188
Europe (including FSU)	644	302	279	239
Asia	11	116	176	176
Africa	9	33	60	61
Australasia	32	36	68	70
Latin America	2	3	8	8
Total	**1,119**	**690**	**772**	**689**

Source:
1 Zimmermann, E. W., *World Resources and Industries*, Harper and Row, New York, 1951
2 Grossling, B. F., *World Coal Resources*, Financial Times Business Information Ltd, London, 1979
3 *BP Review of Energy Statistics*, 1997
4 *BP Review of Energy Statistics*, 2001

The realities for the world coal industry since 1976 have proved to be radically at odds with such forecasts (Gordon, 1987; Grossling, 1981). Global markets for coal expanded annually for only another 14 years until 1989. For the rest of the century they then stagnated at 2.1–2.3Gtoe a year. Cumulative use from 1976 to 2000 barely reached 50Gtoe (75 x 10^9 tons of coal) – only about 75% of what had been so confidently expected.

Meanwhile, in spite of the significant downgrading in the estimates of the world's ultimately recoverable coal resources, declarations of economically viable reserves decreased only modestly to a little under 700 x 10^9 tons (515Gtoe) by 2000. These still indicate a nominal reserves-to-production ratio of over 300 years.

Globally, therefore, given an annual rate of expansion of coal use during the 21st century which begins at 1.5% a year in the period 2000–09 and then gradually falls to less than 0.5% a year from the 2050s, there can be no relevant issue of scarcity. Cumulative coal use during the 21st century would be only just under 460Gtoe (690 x 10^9 tons of coal). This is equal to present proven coal reserves (see Table 2.2), so that only relatively modest ongoing additions to reserves from the ±5,500 x 10^9 tons of coal (3,600Gtoe) in the world's resource base are required to sustain the industry over the century. Indeed, with coal production of 9.6 x 10^9 tons (6.4Gtoe) in 2100, it would be necessary to add less than 1000 x 10^9 tons to proven reserves over the century in order to secure a nominal 100-year reserves-to-production ratio by 2100. The inability of the industry to convert only 18% of the presently remaining resource base into proven reserves over 100 years is, quite simply, inconceivable. A slowly growing world coal industry on the scale hypothesised above would seem well able, under conditions of real prices maintained at present levels and in the context of growing knowledge and advancing technology, to achieve an average annual addition to proven reserves which is little more than the amount which will be produced and used each year over the century.

Environmental Constraints

The hypothesised annual growth rates of coal production over the 21st century are thus essentially modest, when compared with the significantly higher rates of exploitation which the resource base could sustain. The restraint reflects, initially, the continuing competition from hydrocarbons for at least the first half of the century (see Chapters 3 and 4), whereby coal's share of the total carbon fuel supply continues to remain below 30% until 2050. Thereafter, the effects of the expansion of the availability of alternative (renewable) energies principally for power generation will constrain the use of coal. Nevertheless, it eventually seems likely to regain some of the ground it lost in preceding decades to oil and gas. Its contribution to carbon fuel supplies by 2100 could well be back above 30%. Throughout the period, however, coal production potential will be constrained by opposition to its increasing use on environmental grounds, most notably relating to localised atmospheric pollution, but also for fears of global warming from its heavy CO_2 emissions (Grübler, 1999).

The still unresolved problems – after two decades of effort – associated with the commercialisation of 'clean coal technology', whereby emissions from coal burning are virtually eliminated, seem likely to make companies and governments reluctant to finance its development on a large scale (IEA, 1997). Other options for investments have already become relatively more attractive, notably those arising from major advances in gas-use technology. Such developments will persist for the next few decades and will greatly stimulate gas supplies (see Chapter 4) until after 2080. Thereafter, the direct conversion of solar power or other renewable technologies will dominate. It is not without significance that the late 20th century heavy investments of the major oil companies in international coal exploration and exploitation are now being replaced by investments in alternative energies; most notably for the production of electricity from benign solar sources such as wind power. These investments will lead to direct competition in that most important of all markets for coal – the generation of electricity.

Thus, the late 20th century widely held expectation that coal would become the main source of energy of the 21st century (Grübler, 1999; Marchetti, 1978), as it had previously been in the principal source in the 19th, is now being undermined at an accelerating rate. Although the global coal industry will more than double in size in absolute terms between 2000 and 2100, its share of world energy supplies will fall from its current 25% to 20% by 2050 and to only 17% in 2100.

Regional Considerations

Although coal occurrences are widely distributed around the world, with an estimated 2,100 coal-bearing basins (Matveev, 1976) of which 221 had original geological reserves of over 500m tons, the significance of coal as an economic resource is now restricted to just 27 countries – of which ten are involved primarily as consumers of imported coal.

Very large reserves – and reserves potential – are restricted to ten countries (see Table 2.3) with the first two, the US and Russia, sharing 38.1% of proven reserves. These ten countries account for 92.3% of the world's reserves and are broadly distributed across the continents, except for South America, where Colombia, the Latin American country richest in coal, has only 0.7% of world proven reserves.

Table 2.3: Rank ordering of leading countries' proven coal reserves compared with share of world production, 2000

	Share of world reserves	% Cumulative share of world reserves	Share of world production
United States	24.4	24.4	26.5
Russia	13.7	38.1	5.4
China	11.7	49.8	23.5
India	11.1	60.9	7.4
Australia	8.4	69.3	7.2
South Africa	6.5	75.8	5.9
Germany	6.0	81.8	2.7
Kazakhstan	4.4	86.2	1.8
Ukraine	3.3	89.5	2.0
Poland	2.8	92.3	3.4

Source: *BP Statistical Review of World Energy*, 2001 (with sub-bituminous coal and lignite reserves adjusted to hard coal equivalent).

Table 2.4: Rank ordering of leading coal exporters, 2000

	Net exports (mtoe)	Share of coal production exported (%)
Australia	108.3	69.5
South Africa	44.6	64.7
United States*	34.5	5.8
Indonesia	33.7	75.9
Colombia	22.6	91.1
Kazakhstan	15.3	39.7
China	8.1	1.6
Canada	7.6	20.5
Russia	5.4	4.6
Czech Rep.	3.9	15.5
Ukraine	3.3	7.8

Source: *BP Statistical Review of World Energy*, 1998
* Data for the US is for 2001 as production/exports were disrupted by a miners' strike in 2000

Only eleven countries have net exports of more than 4.5m tons of coal (= 3.0 million tons oil equivalent) a year (see Table 2.4), but this group notably excludes two of the top seven rank-ordered countries by reserves; viz.

- Germany, the declared reserves of which ought probably to be significantly discounted as most of the county's coal is either uneconomic to produce, or too polluting to use in the context of a stringent emissions policy. Indeed, as shown in Table 2.5, Germany is the world's third largest importing country, given the wide differential between the price of imported coal and the much higher cost of producing its indigenous coal;
- India, where the large-scale exploitation of its coal reserves is now dedicated entirely to the demands of the country's own energy economy. It has, indeed, recently become a net importer with a slowly growing need for foreign coal for use in areas of the country which are remote from the country's own coal production.

These two anomalies show up in Table 2.5 in which Germany and India feature as the 3rd and 6th largest net importers of coal.

Table 2.5: Rank ordering of leading coal importers, 2000

	Net imports (mtoe)	% Share of imported coal in total net energy imports
Japan	97.1	23.2
Taiwan	36.4	36.5
Germany	28.2	13.4
South Korea	27.2	18.4
UK	17.9	100.0
India	14.4	18.9
Spain	13.4	14.3
Italy	13.0	9.0
France	11.7	8.4
Brazil	10.7	30.8
Thailand	8.8	30.6
Netherlands	8.6	12.3
Belgium	7.6	28.6
Denmark	4.1	28.6
Slovakia	4.0	30.8

Source: Derived from *BP Statistical Review of World Energy*, 2001

Overall, the geographical spread of coal reserves-rich countries provides a continuing high potential for regional supply arrangements which are convenient and relatively low cost (in production and transport terms) for meeting the future needs of the world's leading coal importers. These importing countries divide into two groups; viz.: Western and Central Europe and the Western Pacific Rim.

Western and Central Europe

Present annual import requirements total about 160 million tons (105mtoe). The high, and still rising, costs of most indigenous production imply ongoing increases in imports of coal, either to substitute indigenous coal and/or to meet rising demand. Indeed, imports would double were there to be a substitution of all uneconomic local production by coal imports (Parker, 1994; Radetzki, 1996). But even imports of up to 200mtoe (around 300m tons of coal) would not pose a security of supply problem, taking into account the large reserves of existing suppliers (essentially the US and South Africa). Even if these countries could not supply the required rising volumes, then there are other potential exporters to Europe, such as Colombia and Russia, both of which would be motivated to seek opportunities in the market.

Apart from substituting high-cost indigenous coal supplies to existing markets there must be doubts about the opportunities for growth in the level of coal imports to Europe to serve new markets in the medium term – to at least 2025. This is because of the increasing competition for coal from natural gas, a process which is already well under way. In the longer term there will be competition from alternative, renewable sources of energy which both individual European countries and the European Union are aggressively pursuing. Indeed, the strength of these forces is such that there must be serious doubts as to whether existing – let alone growing – imported coal demands will emerge for all the net importing countries. Belgium, Denmark and the Netherlands, all of which pursued strong pro-imported coal policies for the last two decades of the 20th century, are now moving away from coal because of concerns about its local environmental impacts and, even more so, its high emissions of CO_2. In the UK, market forces have already so strongly favoured gas that the demand for coal in the 1990s declined by almost 50%. This move away from coal is, moreover, already being emulated not only by other western European countries, but also in the previously coal-intensive countries of Central Europe, notably the Czech Republic, Poland and Slovakia (Hoffman, 1985; Odell, 1998b).

We thus conclude that Western and Central Europe's call on coal supplies from elsewhere in the world will, at best, grow slowly to no more than double the present levels by 2050 and will then either stagnate at that higher level or, more likely, retreat slowly from it, as renewable energies emerge as the growth element in the region's emerging energy economy. At worst, coal imports will grow little beyond their present modest level as a consequence of gas competition, and after 2050 they will then decline from that level through competition from renewables. The latter could, incidentally, include the re-expansion of nuclear power based on the commercialisation of 'inherently safe reactors' (Grimston and Beck, 2002).

South East Asia and the Western Pacific Rim
This area has over the last decade become a more formidable importer of coal. As shown in Table 2.5, each of the three main importing countries – Japan, South Korea and Taiwan – imports coal as a significant contributor to total energy imports. Overall, the nations in the region currently import almost 300 million tons (200mtoe) of coal. The strong upward trend in net coal imports established during the 1990s was ameliorated by the late 1990s' economic crisis in the region, but it is now reasserting itself as the setbacks which affected economic progress are overcome. Unlike Europe, the countries in this region do not yet have access to pipelined gas in large volumes and, except for Japan and South Korea, the countries concerned have not developed significant nuclear industries (Grimston and Beck, *ibid*).

Their economies, again excluding Japan, Korea and Taiwan, are also at a lower level of development – including a lower degree of electrification – so that energy use expansion still has some considerable way to go before annual rates of growth fall to near zero (as in much of Europe). Finally, the countries do not have the same commitment as Europe to the development and use of renewable energy resources so that increased coal fired electricity generation will be required for at least the first quarter of the 21st century (Chow, 2003).

Overall, the total requirements of the region for imported coal will more than double by 2020. Thereafter, however, infrastructure capable of bringing large volumes of natural gas to the markets concerned will be fully in place, so that competition from gas will restrict the rate of expansion in the use of coal. Thus beyond 2020 the increase in coal use will be moderated by gas substitution and, in addition, by the introduction of renewable energy technologies. To judge from historical antecedents in other sectors of the energy economy, the region's expansion of renewable energy production will then be fast and extensive. Thus, early in the second half of the 21st century coal use and imports will reach a peak. By that time imports of coal seem likely to be in excess of 750 million tons (500mtoe) a year.

In the meantime, however, the region's significant proven coal reserves and considerable additional potential will continue to expand. Australia and India together will still have over 100 gigatons of proven hard coal equivalent reserves – equal to over 130 years' supply at the mid-21st century indicated level of imports for the Western Pacific Rim countries. Even excluding the sub-bituminous and lignite reserves from the equation still leaves over 100 years' supply from already known hard-coal resources within the region.

The coal reserves/supplies prospects of the region are, moreover, significantly supplemented by the very large coal resources of China. As shown in Table 2.3, it is third in the world in terms of reserves, with 11.7% of the global total. These reserves, however, already sustain some 25% of the world's production of coal; almost the equal of that of the United States, even though its reserves are less than half those of the US. Although China remains self-sufficient, its reserves-to-production ratio of 118 years is now relatively modest compared with the reserves wealth of the other top ten countries (notably the US with an R/P ratio of 209 years). China's ultimately recoverable reserves are, however, generally estimated to be up to 10 times greater than its proven reserves, but many of these are inaccessible (in the north-west of the country), while most of the rest remain untested.

Whether or not China emerges as anything more than a minor exporter of 15 to 30 million tons per year to the Western Pacific rim region seems to depend, in the short to medium term, more on its policy towards the exploitation of coal, rather than on coal use in China itself. Its high level of coal use (it still accounts for just under two-thirds of the country's total energy use) is now recognised not only as seriously overloading its transport infrastructure (because of the heavy use of the country's railway system for coal traffic), but also as the cause of horrendous pollution, most notably in and around the largest cities. Efforts to curb the expansion of coal use within the country could lead to an expanded export capability from the present modest level indicated above. Whilst a degree of dependence on increased coal imports from China would be acceptable as a means of import diversification by Japan, Korea and other western Pacific countries, there would certainly be political concern if, in the long-term, a much higher degree of dependence on China's coal reserves and their exploitation were to emerge.

The Rest of the World
Elsewhere in the world the issue of long-term access to coal reserves has not arisen – and seems unlikely to do so. The fifteen countries which remain upwards of 20% coal-dependent in their total energy use (see Table 2.6), are with two exceptions (Taiwan and South Korea) so dependent simply because they have large reserves of coal. Over the short to medium term their coal dependence seems, in most cases, as more likely to decline than to increase. This will be either because they find alternative, albeit imported, sources of energy to be more economic, or because they recognise that their environments are too heavily polluted as a consequence of too much coal use. For a number of the countries concerned, viz. Germany, Greece, South Korea, Taiwan and Australia, their 1997 Kyoto Treaty commitments to reduce CO_2 emissions may well require them in the near future to pursue policies which effectively limit, or even reduce, their coal use more quickly over the next decade.

In other parts of the world – most notably Latin America, most of Africa and the Middle East – the absence of known coal reserves or the ability to exploit those which do exist, led to a reliance on other energy sources (e.g. hydroelectricity and oil or gas). There are near-zero prospects of these countries deliberately pursuing policies which aim to increase their use of coal. Where other energy resources are indigenous to these countries, their exploitation will be encouraged, but where dependence on imports is inevitable, then their current preference for importing oil or gas will continue to be exercised. Or changed to a preference for natural gas, as in most of Latin America, where gas transportation systems are being developed to connect gas-rich areas within the continent with the continent's city regions which are highly energy intensive.

As will be shown in the following chapters, such continuing hydrocarbons-based development, involving increases in the volumes required, is generally possible over, at least, the first half of the 21st century. If scarcities eventually develop so that prices rise to unacceptable levels, then progress already made elsewhere, notably in the industrialised countries, in the commercialisation of renewable energy sources seems more likely than coal imports to provide the alternative means to energise the development process. If so, then the geographical expansion of the world's markets for coal will be increasingly constrained. There is even likely to be a continuing reduction in the fifteen presently coal-dependent countries (shown in Table 2.6), as a "natural succession" to the exclusion over the last decade of 10 other countries which were still at least 20% dependent on coal in 1990 (including Russia, the UK, Denmark, Ireland and Spain).

Table 2.6: Rank ordering of countries at least 20% coal-dependent in 2000

	Contribution of coal to total energy supply (%)
South Africa	75.6
Poland	65.2
China	61.4
Kazakhstan	57.3
India	54.7
Czech Republic	52.0
Australia	43.9
Taiwan	33.8
Turkey	30.3
Ukraine	29.3
Greece	28.9
Hungary	28.8
Germany	25.7
US	24.7
South Korea	22.5

Source: *BP Statistical Review of World Energy*, 2001

Possible New Developments

Overall, the expansion prospects for coal face a set of barriers. Any major resurgence of coal in terms of its contribution to the 21st century energy supply – and thus of making use of the effectively near-limitless resources available (in relation to demand) – depends on breakthroughs in the technologies of production and use.

On the supply side, such a breakthrough must relate to the successful introduction of the underground gasification of coal and the supply of its energy in the form of a gas – rather than as the solid form in which it has been used to date. The considerable number of underground gasification experiments (in many of the major coal producing countries) has so far failed to sustain the required continuous deliverability of coal-energy in this form (IEA, 1997). And even if it were successful, it would produce such low calorific gas as to be able to serve only those markets in close proximity to the supply. Such a development would thus not be sufficient in itself to globalise the market for coal. Alternatively, countries with exploitable coal reserves and a propensity to use them could do so, without compromising their commitments to the Kyoto Treaty in terms of required reductions in CO_2 emissions, by ensuring the sequestration of the CO_2 produced by coal combustion. This technology has yet to be developed as an economic proposition, but rapid progress is under way (Freund, 2002; Pachin, 2003).

On the demand side, market expansion requires the massive development of combined cycle/coal gasification plants, whereby both electricity and heat from the plants can be delivered to users. A few such processing units are operational, but both technical problems and high costs, given the capital-intensive nature of the plants, continue to inhibit their wide-scale use (Parker, 1994). In the medium term, the applicability of the technical possibilities will remain highly location-specific, rather than generally possible. For the longer term – say from 2025 onwards – their development could become economically attractive in areas where depleted gas resources leave a gas pipeline infrastructure available for moving the gas produced from coal to customers hitherto dependent on natural gas. Even in these cases, however, there will be competition from the prospective hydrogen-orientated energy system for the follow-up use of such available pipeline transport and distribution infrastructure (Hoffman, 2001; Rifkin, 2002).

Should oil products from crude oil and from natural gas become inadequate to serve market demands – in the short to medium term in specific locations and in the longer term more generally – then coal-to-oil-products conversion plants, either stand alone or in the context of more comprehensively designed 'energy-plexes' (Odell, 1974), could be located in coal-rich areas yielding coals of appropriate qualities. The evolution of this potentially important future use for additional volumes of the world's extensive coal resource base will depend, first, on constraints on the long-term future supply potential of crude oil and natural gas and, second, on the degree to which traditional petroleum fuels remain acceptable for vehicle propulsion. The hitherto widely accepted potentially 'seamless' switch from oil to coal for the derivation of gasoline and diesel oil from oil-from-coal plants (as developed by SASOL in South Africa) for the world's increasingly large numbers of motor vehicles now seems less likely, given the environmental and emissions constraints on the future use of such vehicles, compared with the direct or indirect use of natural gas as a medium-term alternative (Douaud, 2002) and the use of hydrogen in the longer term as another (European Commission, 2003).

References

Chow, L., (Ed.) (2003), "Themes in current Asian energy," *Energy Policy*, Vol.31.11.

Douaud, A. (2002), "Oil, gas, hydrogen and electricity; energies of the future for transport," *Proceedings of the 17th World Petroleum Congress*, Rio de Janeiro, Vol.1, pp.303–18.

European Commission (2003), *Towards a Hydrogen-based Energy Economy*, Brussels.

Freund, P. (2002), "Technology for avoiding CO_2 emissions," *Proceedings of the 17th World Petroleum Congress*, Rio de Janeiro, Vol.5, pp.11–21.

Gordon, R.L. (1987), *World Coal: Economics, Policies and Prospects*, Cambridge, Cambridge University Press.

Green, R.P. and Gallagher, J.M. (Eds.) (1980), *Future Coal Prospects: Country and Regional Assessments*, Cambridge, Mass., Ballinger.

Grimston, M.C. and Beck, P. (2002), *Double or Quits: the Global Future of Nuclear Power*, London, R.I.I.A.

Grossling, B.F. (1981), *World Coal Resources*, 2nd Edition, London, Financial Times Business.

Grübler, A. et al (1999), "Dynamics of energy technologies and global change," *Energy Policy*, Vol.27, pp.247–280.

Hoffman, G.D. (1985), *The European Energy Challenge; East and West*, Durham, NC, Duke University Press.

Hoffman, P. (2001), *Tomorrow's Energy: Hydrogen, Fuel Cells and the Prospects for a Cleaner Planet*, Cambridge, Mass, Cambridge University Press.

I.E.A. (1997), *Energy Technologies for the 21st Century*, Paris, OECD.

Manners, G. (1971), *The Geography of Energy*, London, Hutchinson.

Marchetti, C. (1978), *Energy Systems, the Broader Context*, Laxenburg, I.I.A.S.A.

Matveev, A.K. (1976), "Distribution and resources of world coal," *Proceedings of the 1st International Coal Exploration Symposium*, London, pp.77–88.

Muir, W. (Ed.) (1976), "Coal Exploration." *Proceedings of the 1st International Coal Exploration Symposium*, San Francisco, Miller Freeman Publications.

Odell, P.R. (1974), *Energy; Needs and Resources*, Basingstoke, MacMillan Education.

Odell, P.R. (1998b), "Energy, resources and choices" in Pinder, D. (Ed.), *The New Europe; Economy, Society and Environment*, Chichester, J. Wiley and Sons Ltd.

Pachin, J.C. (2003), "Coal in Europe: implications of dismantled subsidies," *Energy Policy*, Vol.23.6, pp.481–560.

Parker, M. (1994), *The Politics of Coal's Decline; the Industry in Western Europe,* London, Royal Institute of International Affairs.

Radetzki, M. (1996), 'Fossil fuels will not run out,' *Journal of Mineral Policy,* Vol.12, No.2, pp.26–30.

Rifkin, J. (2002), *The Hydrogen Economy*, London, Penguin Books.

Schurr, S. and Netschert, B. (1977), *Energy in the American Economy, 1850–1975*, Baltimore, The Johns Hopkins University Press.

Chapter 3: Oil's Long Term Future; 85% yet to be Exploited

The Reserves' Discovery and Appreciation Process
The future availability of oil has long been a recurring issue both for the industry and with government policymakers (Williamson, 1963). As long as it remained a relatively unimportant energy source, with its use on a significant scale restricted to a relatively small number of countries (most notably the US), as was the case until after the second World War, a fear of global scarcity was not a real issue. For the few mainly oil using countries before 1939 their concern was related to a perceived scarcity of indigenous supplies.

Post-1945, however, in the context of oil rapidly becoming the most important source of global energy needs – with oil eventually supplying over half of the world's total energy use by 1958 – a widely held perception of the idea that the world would be unable to continue to run on oil for the rest of the 20th century quickly emerged (Odell, 1964). This view grew in strength in the 1960s as oil use expanded at over 7% per annum and was powerfully set out in an article by no less than the chief geologist of British Petroleum in the very early 1970s (Warman, 1972). Though the fundamental validity of the analysis which produced this result was challenged (Odell, 1973), pessimism on oil's future prospects persisted. It was eventually expressed most emphatically and succinctly in a study, subsequently published with a high profile launch by BP, under the title *Oil Crisis… Again?* (BP, 1979). That study purported to show that world oil production – outside the Soviet bloc – would necessarily have to peak in 1985 – only six years after publication, as shown in Figure 3.1. Other companies and institutions concerned with energy forecasting judged the turning point for peak global oil consumption to be around the late 1990s or the early 21st century (Shell, 1979; Grenon, 1979). A total of no fewer than twelve pessimistic studies of oil's prospects at that time were analysed in detail in a research project at the Centre for International Energy Studies at Erasmus University, Rotterdam (Odell and Rosing, 1980/1983). This showed that the fears for an impending near-future scarcity of oil were based on totally inappropriate parameters; first of a world fully explored for oil and, second, as a world in which 'the end of history' was thought to have been reached with respect to the assumed near-future end of the processes of increasing knowledge and advancing technology in the oil industry. These supply-side absolutes were then, moreover, combined with another absolute belief, viz. the absurd notion that oil had a perfectly price inelastic supply curve.

Needless to say, developments in the real world soon undermined all the component parts of these scaremongering hypotheses. By 1979 oil demand growth came to a standstill as the impact of the first oil price shock was fully felt.

Thereafter, the use of oil fell year by year to a low in 1983 (when it was almost 15% under its historic peak reached in 1979). It then took nine years to 1992 for global oil use to recover to its 1979 level. Thereafter, even over the final eight years of the 20th century, growth in oil use remained relatively weak with an increase of only 11% (an annual average of 1.2%); and over the 21 years from 1979 by a mere 11.2% (= 0.5% per annum). Cumulatively, the use of oil over the last 30 years of the 20th century totalled about 90Gtoe – rather than the 250Gtoe which was so confidently forecast by the industry in the early 1970s (Warman, 1972).

On the supply side, the growth in reserves from new discoveries – and, even more importantly, from the appreciation of reserves in fields long since discovered – has run quickly ahead of oil used. The data are shown in Table 3.1. Almost 1,200 billion barrels of oil (165Gtoe) were added to proven reserves between 1971 and 2000. Over the same period only 682 billion barrels (92Gtoe) were consumed. From these data, one can argue for a world which over the last 30 years of the 20th century since 1971 was 'running into oil' rather 'out of it', as so widely forecast in the 1970s.

Thus starting with the situation at the end of 2000 as the baseline for a study of the prospects for oil in the 21st century, 1028 billion barrels (140Gtoe) of proven reserves were then available; not only able to satisfy the year 2001 volume of demand of 27.5 billion barrels (3.5Gtoe), but also another 36 years of oil use at the same level of annual production. Even were oil use to grow at 2% a year from the

Oil's Long Term Future; 85% Yet to be Exploited

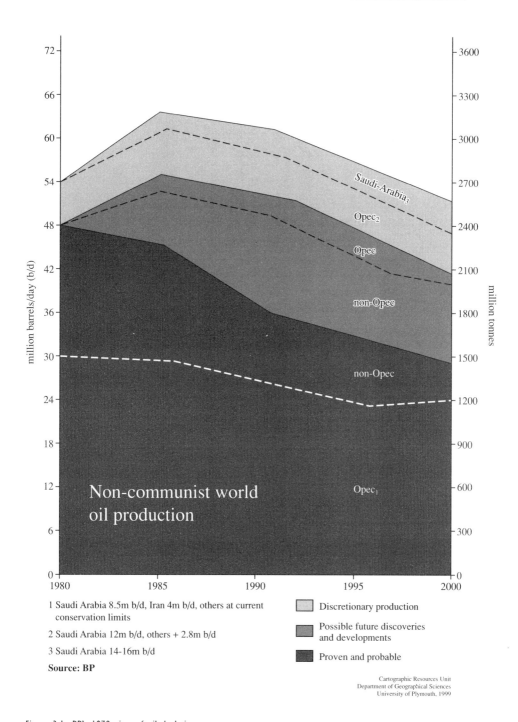

Figure 3.1 BP's 1979 view of oil depletion
Source: BP

Table 3.1: Proven reserves, oil production, gross and net growth in reserves and reserves-to-production ratios, 1971–2000

	Proven reserves at start of year	Production of oil in year	Gross additions to reserves	Net growth or decline in reserves	Reserves-to-production ratio
		(billion barrels)			(years)
1971	521	18.4	38	20	28.3
1972	542	19.4	54	35	27.9
1973	577	21.2	35	14	27.2
1974	591	21.2	32	11	27.9
1975	602	20.2	31	11	28.4
1976	613	21.9	4	−18	30.3
1977	595	22.6	16	−7	27.2
1978	588	22.9	45	22	26.0
1979	610	23.7	22	−2	26.6
1980	608	22.8	34	11	25.7
1981	619	21.3	67	46	27.1
1982	665	20.1	30	10	31.2
1983	675	20.0	21	1	33.6
1984	676	21.1	44	23	33.8
1985	699	20.5	30	9	33.1
1986	708	21.4	67	45	34.5
1987	753	21.9	129	107	35.2
1988	860	22.8	83	60	39.3
1989	920	23.5	87	63	40.3
1990	983	23.8	26	2	41.8
1991	985	23.7	65	41	41.4
1992	1,026	23.9	46	22	43.3
1993	1,048	23.7	31	7	43.9
1994	1,055	24.0	29	5	44.5
1995	1,060	24.3	48	24	44.2
1996	1,084	25.5	40	14	44.6
1997	1,098	26.1	−67	−93	42.1
1998	1,005	26.3	31	6	38.2
1999	1,011	26.7	31	4	37.9
2000	1,015	27.3	40	13	37.2
2001	1,028	—	—	—	—
Totals		682	1,189	507	

Source: Development of reserves based on contemporary data from the annual surveys of world oil reserves in the *Oil and Gas Journal*, 1970–2000; *World Oil*, 1971–2000; De Golyer and MacNaughton's *Annual Survey of the Oil Industry*, 1975–1983; production data from the BP *Statistical Review of World Oil/Energy*, 1971–2000.

2000 base, reserves could theoretically supply the oil required for the first quarter of the 21st century. We can, however, confidently predict that the proven reserves declared at the beginning of 2001 will appreciate in volume. Such appreciation has, indeed, been a long continuing process, based first, on frequent reappraisals of reservoirs' potential; second, on the perfectly normal enhancement of geological knowledge as a consequence of production experience leading to extensions to fields; and third, from improving rates of recovery from the oil-in-place in a reservoir, as a result of significant advances in production technologies (Meyer, 1977; Odell, 1994b; Smith and Robinson, 1997; McCabe, 1998).

A conservative view of the likely appreciation of the volume of oil declared proven by 2000 is shown in Figure 3.2. This indicates a reserves gain by 2020 of about 350 billion barrels (4.8Gtoe): a quantity which is sufficient to extend the availability of oil from the world's currently exploited and known fields by the equivalent of about 14 years' supply at the 2000 level of production (Shell, 2001).

Indeed, it can be strongly argued, first, that the advanced technologies now in use for defining or redefining the size and characteristics of oil fields (particularly through the so-called 4D seismic methodology, whereby the dynamic qualities and behaviour of a reservoir under production can be simulated), and second, the new production techniques now developed (most notably horizontal drilling and enhanced oil recovery methods), together serve to add a significant new dimension to the prospects for enhancing the capabilities of the world's existing oil fields to produce additional oil. To date, large investments in these new technologies have largely been restricted to oil producing areas in North America and the North Sea. They have shown significant success, measured in terms of additional oil production already achieved and of potential for future production that would not otherwise have been possible (Smith and Robinson, 1997; Watkins, 2000).

The application of these technologies in other parts of the world is only a matter of the time required to create, first, the demand for the additional oil and, second, the politico-economic conditions in which the necessary investments by companies possessing the relevant expertise can be made. This applies generally, but it is specifically important in respect of the two richest oil regions in the world, viz. the Middle East and the former Soviet Union.

In the Middle East, the impact of the nationalisation in the 1970s and 1980s of the oil companies working in most of the countries in the

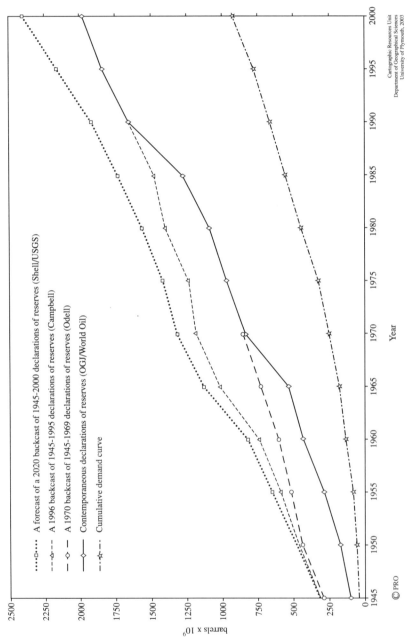

Figure 3.2 The Appreciation of Proven Reserves of Conventional Oil compared with Cumulative Demand, 1945–2000

region, combined with the subsequent financial, managerial and/or technical limitations of most of the state entities to undertake the necessary exploration and exploitation work, has resulted in much of the upstream oil industry in the world's most significant oil producing region now being out of date. An appreciation of the need for radical change is gradually taking place so that agreements to update the industry are now emerging. When the proposed and potential joint ventures between many international oil companies and the various state corporations in Iran, Kuwait, Saudi Arabia, Iraq and elsewhere in the region become operational during the first decade of the 21st century, they will lead to a very significant enhancement of the region's reserves and of its production potential (Odell, 1997a; Centre for Global Energy Studies, 2001).

Similarly, the oil industries of Russia and of other former Soviet republics await the application of new technologies and methodologies through joint venture structures with international oil companies which are financially capable of undertaking the work. As they are gradually put in place, the productivity of the oil industries of countries in Russia, Ukraine, Azerbaijan, Kazakhstan and Turkmenistan seem likely to be revolutionised (Krylov, 1997; Khartukov, 1997; Considine and Kerr, 2002). Indeed, by 2002, the proven reserves of these countries have already increased by almost 40% from their levels of a decade earlier.

What Oil Crisis?

It is, in part, these politico-economic components in the evolution of the prospectivity of world oil that are significant in undermining the validity of the renewed efforts by some to claim that another oil supply crisis is pending (Campbell, 1997 and 2003; Campbell and Laherrère, 1998; Laherrère, 1997 and 2003). Once again, as with their predecessors such as Warman (1972) and Hols (1972) in the 1970s, their forecasts of a near-future peak in global oil production (by 2005/6) fail to recognise the dynamics of the processes whereby oil reserves and production evolve: and they equally avoid the central role played by both economics and politics in equilibriating the markets (Lynch, 1997). Such irrational warnings of an early 21st century oil scarcity should thus be ignored, particularly as one recalls the huge costs that were imposed on the world economy by the earlier acceptance by many energy and economic policymakers of the 1970s' prognostications of oil scarcity (Adelman, 1993; Odell, 2001). That episode was responsible in large part not only for the much higher oil prices and the economic and social problems that these caused, but also for the very large non-viable investments that were made in alternative energy production systems and in the exploitation of higher cost energy reserves (McCabe, 1998; Odell, 1998). The world at the present time can ill afford an unnecessary repetition of that near disastrous set of events.

The world's already proven reserves of oil – and the processes whereby they evolve – thus totally eliminate any significant up-side restraint on the development of production for the first quarter of the 21st century, given the maintenance of an inter-quartile price range for oil at the $18–21/bbl (in $ of 2000) level which emerged over the near-20 year period since the oil price collapse in 1986. On the contrary, any up-side constraint on supply will, for at least the first 20 years of the 21st century, be imposed by slow demand growth which, as shown above, has already been modest for the past 20 years. The author has already indicated in Chapter 1 why such slow growth in energy use in general is likely to continue. Oil demand will, in particular, be constrained, given the increasing competition it has to face from natural gas in many markets across much of the world in the period to 2020 (see Chapter 4).

Beyond that date – and thus also beyond the importance of currently proven reserves – and their evolution through appreciation based on increasing knowledge and technological progress – one has to move into the more uncertain issue of the size of the ultimate world's oil resource base and its prospective exploitation. This has been described as 'the unknown, the unknowable and the unimportant' (Adelman, 1993), a description which in purely economic terms is entirely apt. In a competitive market with many active players, what is demanded is supplied (produced), providing the price is high enough to sustain profitable business. Moreover, demand also has to be anticipated by investments which are made in finding and developing reserves so that, given a long lead-time for this process, the market can be served. As shown above, the industry has responded in exactly this way over the past 30 years, in spite of the pessimism over the long-term future of oil in the early 1970s. This response led to the creation by the late 1980s of a reserves-to-production ratio of almost 40 years, as shown in Table 3.1. In this context, it is thus not surprising that there was only one year since 1979 in which the industry's exploration and development activities did not lead to the full replenishment of the stock of reserves, viz. in 1997. This was the year when a number of countries' reserves declarations were downgraded by *World Oil* because of doubts as to the recoverability of oil in the context of a sharp downturn in prices (see Figure 1.3). This clearly indicates that the normal economic process of stock renewal is working effectively. Any serious concern over the rate of conversion of the world's oil resources to reserves would be justified only in the event of a run of consecutive years (a minimum of, say, five or six) in which annual production exceeds the gross additions to reserves.

There is an argument, most recently used by Campbell (1997), that annual additions to reserves (comprising both new discoveries and the appreciation of reserves in previously discovered fields) should not be taken to indicate replacement or replenishment of the reserves stock. Such replenishment should, he claims, be judged only against discoveries of entirely new reserves, while additions to reserves arising from the re-valuations of earlier discoveries should be dated back to the year of the initial discovery of the field concerned. But this is not a robust argument. As far as the oil economy is concerned, the "why and the wherefore" of the development of reserves are immaterial. It is the *fact* of their occurrence and of their declared availability at a particular time to supply the market that is of the essence in terms of balancing future supply against future demand.

This makes the nature of the time-series which Campbell (*ibid.*) has produced, in which data on the increases in reserves from discovered fields are backdated to the year of discovery, invalid for purposes of forecasting supply. Under this procedure, the more recently discovered fields have had less time to go through the normal process of appreciation than fields discovered many years – or even decades – ago, so that comparisons between the recoverable reserves in long-since discovered fields and those in recently discovered fields are rendered invalid. The backdating of reserves with hindsight in the context of newly developed technologies of reserves' assessments and recoverability, coupled with significantly changed production cost and market prices is simply inappropriate to the contemporaneous economic evaluation of oil exploitation: it makes the past look more attractive than it really was to the economic decision makers of the time; while the present is made to appear less attractive.

For example, as shown in Figure 3.2, the prospects for the industry in 1960 depended on the then indicated existence of only about 200 billion barrels (2.7Gtoe) of proven reserves at a time when some 135 billion barrels (1.9Gtoe) had already been used, so that the volume of reserves then declared appeared to be modest compared with the annual rate of use. But ten years later in 1970 it was shown that 300 billion barrels (4.0Gtoe) of recoverable oil had existed in the fields discovered by 1960. Then, with the passage of another 25 years to 1995, the graph shows that the recoverable reserves from the population of fields in 1960 had increased still further to about 430 billion barrels (5.4Gtoe). Thus, the feared pending oil scarcity suggested by some pessimists in the early 1960s, predicated on the basis of the argument, "almost half of the world's known oil reserves had already been used," was entirely groundless (Odell, 1964).

Likewise in 1995, when the contemporary scaremongering of oil scarcity again re-emerged (Campbell and Laherrère, 1998), it was argued by the new "Jeremiahs" that the approximately 1000 billion barrels (13.5Gtoe) of proven reserves were able only to sustain growth in production for another four or five years because by then "half of the world's oil would have been used up", so that the tenuous Hubbert hypothesis, viz. that production is bound to start falling once 50% of known oil has been used, would become applicable (Deffeyes, 2001; Holtberg and Hirsch, 2003). But, that has not happened, simply because cumulative additions to the declared proven reserves in 1995 have virtually kept pace with cumulative oil production. Thus, the 1060 billion barrels of reserves declared in 1995 are already known to have been closer to 1150 billion, even though the process of the appreciation of the reserves declared by 1995 is by no means over. As shown in Figure 3.2, such appreciation is likely to be over 350 billion barrels (4.8Gtoe) by 2020.

Now that the importance of the appreciation over a lengthy period of time of previously discovered reserves has been so clearly demonstrated, we can clearly see why the pessimism in earlier decades over the future of oil was totally unjustified. The 1971 declared economically producible reserves of about 520 billion barrels (7.1Gtoe) are, as shown in Figure 3.2, now known to have exceeded this volume by almost another 500 billion barrels (6.8Gtoe). Contemporary recognition in 1970 of the significance of the phenomenon of reserves appreciation would have eliminated the pessimism that was then so generally expressed (Odell, 1973). It may, indeed, even have inhibited the oil price shock of 1973–4 and, thereafter, the whole gamut of adverse consequences for the international oil industry, in particular, and for the world economy, in general.

The high probability that the year 2000 declaration of over 1000 billion barrels (13.8Gtoe) of proven reserves will be shown by 2020 to have been more than 1350 billion – as a result of the factors set out above – is a highly significant input to the evaluation of the prospects for the conventional oil industry over the 20 year period. Of more emphatic significance is the fact that even without any further discoveries of oil, the peak of conventional annual oil production will not occur during the present decade – as threatened by Campbell (2003) – for supply-side reasons; unless the price of oil collapses, so undermining hitherto profitable production operations. It is falling demand that is more likely to produce a premature peak in global production.

Nevertheless, no matter how plentiful the presently declared reserves of conventional oil are – after due account has been taken of their future appreciation – they still provide only for a finite future of increasing oil production: to the point when steadily increasing demand will have depleted some 50–60% of the calculated, up-dated reserves, in both already-discovered, as well as yet-to-be discovered, fields. When this stage in the evolution of the world's conventional oil industry is eventually reached, the remaining reserves must clearly be 'saved' to serve the market in the subsequent period of declining output. With approximately 900 billion barrels (12.3Gtoe) of oil used by 2000, some 37% of the world's currently-known recovered and recoverable reserves of oil have already been used. Without any more discoveries, approximately 50% will have been depleted by 2011 and about 60% by 2019, on the assumption of an average 1.5% per annum increase in consumption in the meantime. Thus, without a continuing oilfields' discovery process, conventional oil production would have to peak in the late years of the second decade of the 21st century. In other words, in other than strictly economic terms under which an approaching recognition of relative scarcity would lead to a rising long-run supply price and a consequential restraint on demand, continuing discoveries of new oil fields are of the essence in ensuring the ability of the industry to sustain an increasing level of global production post-2020. These prospects are analysed below.

Ultimate Conventional Oil Resources' Depletion, 1940–2140

As Figure 3.3 shows, there have since the 1940s been a large number of estimates of ultimately recoverable oil. At that time the world was thought to have less than 100Gtoe (730 billion barrels). Estimates rose rapidly in the late 1940s and throughout the 1950s and 1960s, as the global oil industry not only expanded geographically, but also increased its intensity of development. Most notable, of course, in this latter respect was the Middle East which was exploited on a large scale for the first time during that period. By 1970 estimates of ultimate global reserves had settled down at around 300Gtoe (2,200 billion barrels) leading to a major fear of scarcity in the context of the widely held belief that this figure represented the ultimate truth on the future availability of oil (Warman, 1972); albeit a view of the limited prospects which was challenged by other observers (Odell, 1973; Styrikovich, 1977; Odell and Rosing, 1980). Since then, however, the hitherto rapid increase in the demand for oil from 1950 to 1973 (averaging 7.5% per annum) fell sharply, so significantly undermining the previously perceived need for large volumes of future supplies. Meanwhile, interest in the ultimate reserves of the Middle East evaporated with the nationalisation of the international oil companies in most of the countries in the region. Instead, a more intensive appraisal of the oil wealth elsewhere in the world became of greater interest – in both the industrialised and the developing countries (Odell, 1981).

As also shown in Figure 3.3, estimates of ultimately recoverable reserves now reach to much higher levels (to over 500Gtoe), although four recent estimates remain at under 300Gtoe. These latter estimates provide quite an extraordinary view of the prospects, given that a total of some 265Gtoe of oil has either already been produced or declared as proven. The observers (defined as the 'Flat-earthers' in the key on Figure 3.3) responsible for these forecasts are, in essence, simply repeating the discredited 1970s belief in the proverbial 'end of history', when almost all of the world's oil was considered to have already been discovered. There are not, they argue, any additional regions of great potential which remain unexplored; nor do they visualise any likelihood of the continuing ability by the oil industry further to enhance the percentage rate of recovery of oil from known reservoirs through continuing technological advances. The views add up to nothing less than a proverbial 'flat-earth theory' in which the sciences and technologies of oil discovery, development and exploitation are at the edge of that world and are about to fall off into oblivion (Hiller, 1997; Campbell and Laherrère, 1998; Deffeyes, K.S., 2001). It is inconceivable that the hypothesis can be correct, given not only the absence of any indicators for the cessation of the industry's exploratory and production research activities (Downey et al, 2001), but also the absence of the critically important economic indicator which would emerge in the context of an impending near future scarcity, viz. a long-term rising real price for the commodity (Adelman, 1993; Lynch, 1997).

Oil's Long Term Future; 85% Yet to be Exploited

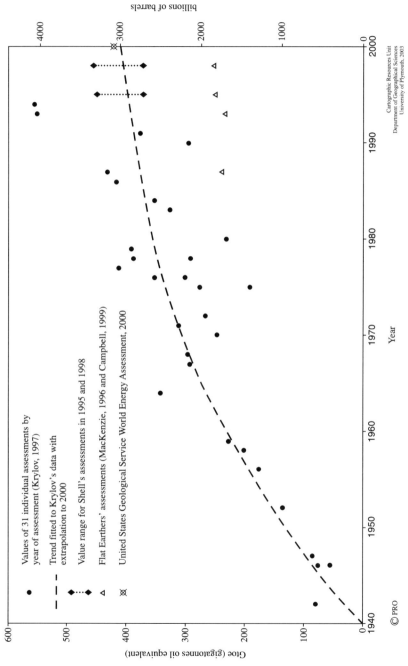

Figure 3.3 Assessments of total world initial oil reserves over the period 1940–2000

Referring back again to Figure 3.3, however, it can be seen that all of the assessments made since 1980 (except, that is, for the four of the 'flat-earthers'), lie well above 300Gtoe (2,200 billion barrels), while four assessments are above 400Gtoe (3,000 billion barrels). This latter figure was already the mid-point of the range presented by Shell in 1995 in its assessment of the world's ultimate reserves of conventional oil (Shell, 1995). This assessment, of course, included the oil that had then already been used (105Gtoe) and the oil that had already been proven (146Gtoe). Shell further reported 'an estimated 500–1,000 billion barrels (70–140Gtoe) of oil yet to be discovered, plus a further 400–500 billion barrels (55–70Gtoe) of oil which are expected to be recoverable from known fields through the wider application of current and new technologies' (*ibid.*). This adds up to a range of ultimately recoverable reserves of 2,700–3,300 billion barrels (365–-450Gtoe). Shell's more recent 1998 and 2001 analyses basically confirmed its previous estimates of ultimate reserves, whilst the United States Geological Service in its massive World Petroleum Assessment exercise involving several years of work by hundreds of oil geologists established a mean value for the world's oil resource base of 3003 billion barrels (420Gtoe) (USGS, 2000; Groenveld, 2002). Thus, the mid-point of Shell's estimates as well as the mean value defined by the USGS assessment, and the mid-point of the highest and lowest of all the other assessments shown in Figure 3.3 is just over 410Gtoe (approximately 3000 billion barrels). This was also the mean value for the world's ultimately recoverable volumes of conventional oil which emerged from analyses of the future of oil made by the author of this book together with colleagues at Erasmus University Rotterdam in the early 1980s (Odell and Rosing, 1980/3).

This figure thus provides a generally and widely accepted figure on the basis of which a full depletion curve for the depletion of the world's conventional oil can be constructed. This curve is shown in Figure 3.4 for the period from 1940 (when only 0.34Gtoe of conventional oil had been produced) to 2140 when the proverbial 'last' economic-to-produce barrel is close to being extracted. The graph shows a peak production year of 2030 at 4.6Gtoe – compared with around 3.6Gtoe in 2000. The further expansion of conventional oil production is thus demonstrated to have some 30 years to run. At that time it will peak at a level about 28% above present production levels. This allows for an average annual growth rate in output of about 1.2% a year until 2020 and, thereafter, at a gradually slowing annual rate of increase over the following decade until peak production is reached in 2030.

Non-conventional Oil Enters the Market

Thirty years is by no means a long time horizon for the future of oil; nor is the indicated annual rate of increase in output very different than that of the last 30 years. The future of oil over the period to 2030

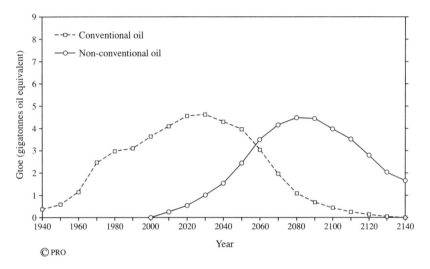

Figure 3.4 Production curves for conventional and non-conventional oil, 1940–2140

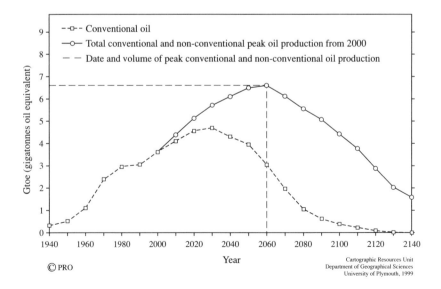

Figure 3.5 The complementary relationship of conventional and non-conventional oil production, 1940–2140

and beyond is not, however, dependent only on the expansion of output from the conventional oil resource base as described and discussed above. It will also involve the production of significant volumes of so-called non-conventional oil – which can be simplistically defined as oil which has to be recovered from habitats other than reservoirs in which oil occurs as a liquid with a viscosity which makes it capable of flowing or being pumped to the surface (Meyer, 1997; Martinez and McMichael, 1997). In terms of both geology and chemistry, the distinction between conventional and non-conventional oil is by no means absolute. Indeed, most of the latter has been converted from the former by degradation, involving significant changes in the chemistry of the oil and, therefore, modification of its physical properties. Moreover, as far as the interest in non-conventional oil from an economic standpoint reflects the ability to derive useful petroleum products – or close substitutes for such products – from it, then no division between conventional and non-conventional oil is strictly necessary. The availability of products from the latter to serve oil markets can be viewed as part of the continuum of a very long-term oil supply process. It is, indeed, no different from the ways in which supply developed in the past; as, for example, in the following developments:

- the extraction of heavier oils from conventional reservoirs as improvements in technology made its production and refining economically viable;
- the winning of oil from offshore fields in the Gulf of Mexico to supplement supplies to the US market previously derived from onshore fields in Texas and Louisiana; or in the evolution of North Sea offshore oil production to substitute for supplies from elsewhere in the world.

These relatively recent changes in supply patterns were 'seamless operations' in the context of the industry's continuing technological process and its ability to organise and finance the development of the new supplies (Odell, 1998). Most oil consumers remained blissfully unaware of these changes in the origin of their supplies. The same will be true in respect of the future switch to non-conventional oil production.

It is interesting to note that the only UN organisation concerned with oil *per se*, viz. the UN Institute for Training and Research Centre for Heavy Crude and Tar Sands (UNITAR), straddles the technical divide between conventional and non-conventional oil. The significance of this is further enhanced by (or perhaps is even a function of) the location of UNITAR in Alberta, Canada, where the more than 50-year-old conventional oil industry has now been complemented by the world's largest existing commercial operations for the recovery of non-conventional oil (from the Athabasca tar sands). These are already producing over 300,000 barrels a day (b/d) of oil products and 500,000 b/d of bitumen (Verbicky, 1998; Lainier, 1998; National Energy Board of Canada, 2000). These operations are, moreover, largely owned by companies whose principal activities elsewhere in Canada and in other countries are concerned with conventional oil (and gas) production. A similar process of diversification by such companies is also under way in Venezuela where, in recent years, the experience of the state owned industry has been extended from the production of heavy oil to the initial production of the reserves of the vast Orinoco oil belt in the form of orimulsion, involving a different and innovative technology (Aalund, 1998). In other words, an equally 'seamless' process in the Venezuelan oil industry has, in effect, already led to the initial stage of diversification of the country's traditional oil industry to non-conventional oil production (Williams, 2003).

The Athabasca tar sands and the oil belt of the Orinoco region are thus self-evidently the sources of non-conventional oil from which conventional oil supplies can initially be complemented. Estimates of the oil in place in the two areas are conservatively put at 4,000 billion barrels, of which up to 15% could be extracted with present technology. This volume of recoverable non-conventional oil is already equal to 40–67% of Shell's estimates of the expected additions to global reserves of conventional oil (see p.48). Even more significant, no less than 178 billion barrels of non-conventional oil in western Canada were formally declared as proven reserves in 2002 – thus making Canada now second only to Saudi Arabia in its oil wealth. Increasing knowledge and improving technology have already led to more than 50% real cost reductions in non-conventional oil production. Thus – in the context of real oil prices which remain at the late 20th century level through the first decade of the 21st century – very large-scale developments in Canada and Venezuela will be under way by 2010. There are, moreover, known extensive occurrences of non-conventional oil in many countries, including Brazil, China, the former Soviet Union, India, Madagascar, the US and Zaire. Their exploitation not only requires conditions which attract large investment funds and technological expertise, but it also implies that there is a requirement for such oil developments in order to meet global or regional demands.

Nevertheless, the continuing absence of any significant motivation for a comprehensive and systematic evaluation of ultimately recoverable reserves of non-conventional oil, in the context of adequate supplies of conventional oil to meet the slowly rising demand until 2020, necessarily undermines the utility of attempting to define the non-conventional resource base of proven, probable and other reserves. Yet this is the most effective basis on which a potential production curve for such oils in the 21st century could be established. In its absence we can do no better than take a deliberately modest figure for future non-conventional oil availability and on that basis define a full-life depletion curve starting from 2000. This exercise is shown in Figure 3.4, for which there is an assumption of a total ultimately recoverable resource base of non-conventional oil of 3,000 billion barrels (500Gtoe). The curve derived from this volume of resources shows a slow build-up to peak production of 4.4Gtoe in 2080 – an output at that date which is of roughly the same magnitude (viz. 4.6Gtoe) as that of the peak for conventional oil some 50 years earlier. Finally, as shown in Figure 3.5, when the depletion curve of non-conventional oil is added to the depletion curve for conventional oil, non-conventional supply is seen, first, to take over as the more important growth element in the total oil supply curve by 2020 and, second, to push the combined peak production of oil for another 30 years into the future – to 2060: to a joint production level of 6.6Gtoe, almost twice that of global oil output in 2000. Thereafter, this combined output of conventional and non-conventional oil enters a long period of decline. Even so, in 2100 the combination of the outputs of conventional and non-conventional oil still supports an oil industry which is approximately 28% larger than the industry in 2000.

In other words, we argue with confidence that large volumes of oil will continue to be offered to the global energy market throughout the 21st century, but that the expansion of the industry will likely cease by about 2060. Thereafter, given the assumptions made above of reserves' availability, a slow decline will necessarily ensue. It seems more likely than not, however, that this supply-side limitation will be subsumed within a somewhat sharper decline engendered by a falling demand for oil, as natural gas and renewables substitute it in an increasing number of end-uses.

The supply of oil has, as analysed above, been divided formally into conventional and non-conventional components. We have, however, previously argued that oil will, of course, be supplied to consumers in future without specific reference to its origin. Indeed, the origin of available oil will from time to time and from place to place be variable, dependent on all the factors that each of the many suppliers of oil has to take into account when determining their supply schedules in the light of changing circumstances. As shown in Figure 3.5, supplies of conventional and non-conventional oil can be viewed as complementary for the whole of the 21st century, but more especially so after 2030 when non-conventional oil production grows to exceed 1Gtoe per year (equal to approximately 20 million b/d). It will, however, be the 2050s before non-conventional oil becomes the more important source of supply.

Nevertheless, over the century, oil's contribution to the total hydrocarbon supply will fall progressively from a 63.8% share in 2000 to only 29% in 2100 (see Table 3.2). This will, in part, reflect a resource base restraint, but, in greater part, it is more likely to indicate a demand-constraint as the global hydrocarbon industry in the 21st century increasingly turns its attention to – and makes more of its investments in – the supply of natural gas, partly for purely economic reasons and partly for environmental reasons. Under those circumstances, the world will not be running out of oil, or even out of the ability to expand supply beyond the limits shown above. Oil could instead be running out of markets in the face of increasing competition from gas, so that, as shown in Figure 3.6, the contribution of gas to the total hydrocarbon supply will already exceed that of oil by the late 2030s. Soon after 2060, as shown in Table 3.3 and Figure 3.7, oil's cumulative contribution in the 21st century to the global hydrocarbon supply will have fallen to less than 50%. The steadily evolving role of natural gas to become the most significant contributor to the global hydrocarbon supply in the 21st century is presented and discussed in Chapter 4.

Table 3.2: The changing contributions of oil to the total supply of hydrocarbons, 2000–2050 and 2100

	Total oil and gas supply (Gtoe)	Total oil supply (Gtoe)	Oil's share of the total (%)
2000	5.8	3.7	63.8
2010	7.1	4.3	60.1
2020	8.6	5.1	59.3
2030	10.5	5.6	53.3
2040	12.5	6.2	49.6
2050	14.1	6.5	46.1
	↓	↓	↓
2100	15.5	4.5	29.0

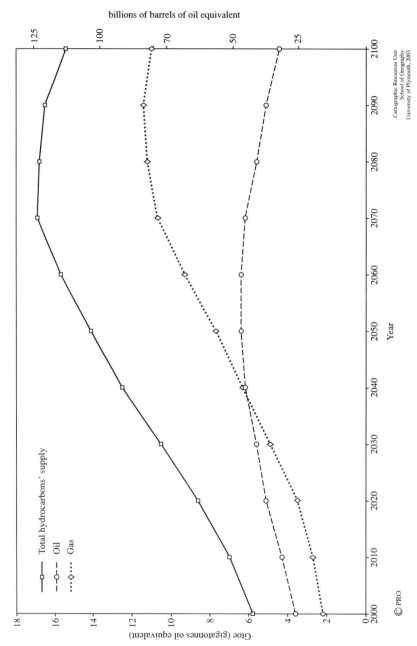

Figure 3.6 The changing contributions of oil and gas supplies by decade in the 21st century

Oil's Long Term Future; 85% Yet to be Exploited

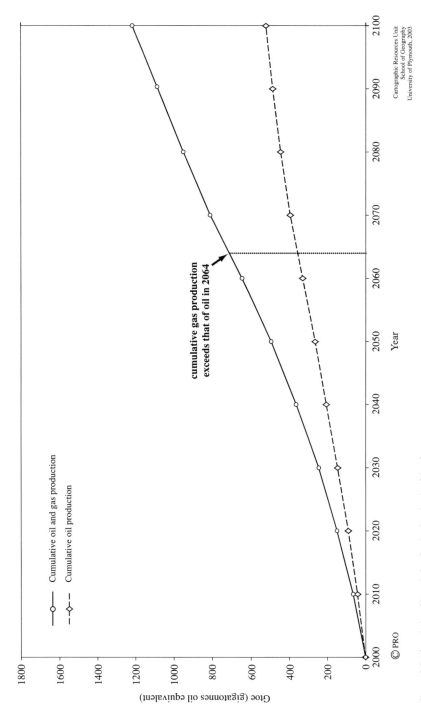

Figure 3.7 Cumulative Oil and Gas Production in the 21st Century

Table 3.3: The cumulative contributions of oil and natural gas to the energy supply in the 21st century

Period	Cumulative oil and gas (Gtoe)	Cumulative oil (Gtoe)	Oil's share of cumulative total (%)	Oil's share per decade indicated (%)
Pre-21st Century	176	120	68.2	–
2001–2010	65	40	61.4	61.4 (01–10)
2001–2020	150	91	60.9	60.5 (11–20)
2001–2030	245	145	59.0	56.2 (21–30)
2001–2040	365	206	56.5	51.6 (31–40)
2001–2050	495	266	53.7	46.3 (41–50)
2001–2060	645	330	51.1	42.7 (51–60)
2001–2070	810	393	48.5	38.3 (61–70)
2001–2080	950	443	46.6	34.9 (71–80)
2001–2090	1085	485	44.7	32.0 (81–90)
2001–2100	1215	522	42.9	28.0 (91–00)

Environmental Constraints on Oil's Production, Transport and Use

In spite of the dangers associated with the production and transportation of crude oil in or over fragile environments (as for example, in the tundra regions of northern Russia and North America and along coastlines near to important maritime routes for tankers), there appear to have been few incidents which created environmental problems which then led to anything more serious than temporary and geographically constrained disruptions to scheduled production and flows of oil (El Hinnawi, 1981; National Research Council, 2003). This reflects both the positive economic impact that oil activities produce for oil-rich localities and, even more so, the ability of the oil industry to secure the continuity of operations. The industry has, moreover, in recent years become increasingly aware of, and responsive to, the need for environmentally friendly attitudes and has put both effort and investment into minimising the risk of accidents (Shell, 1998a) through major oil leaks, spills, collisions and fires. When and where these have occurred they have occasionally led to temporary interruptions to local/regional supplies, most notably in the loss of millions of barrels of oil from the damage done to Kuwait's upstream facilities during the Gulf War in 1991. Except in the case of this problem, when there was a notable impact on the oil price for a short period, other past events have not really been significant at much more than the local level – and certainly not at the global level. Continuing guerrilla actions against oil installations in 2003/4 in American-occupied Iraq, seems to be producing another exception.

There have, however, been instances of restraints on oil exploration and exploitation arising from purely environmental issues. These have usually been resolved through ameliorative measures which, while increasing the costs of undertaking particular operations, have in the final analysis allowed the developments to go ahead in an environmentally acceptable manner. One such example was the development of the Wytch Farm oil field located in shallow coastal waters in an environmentally sensitive area off southern England. After a considerable delay arising from the environmental concerns for visually intrusive drilling and production facilities and of possible oil spillages into the shallow waters, it was eventually agreed that the development could go ahead with the drilling of 10km horizontal wells into the oil reservoir from a remote, land-based and acceptable production location. There have been similar requirements to protect a fragile environment in the Dutch Frisian Islands.

Such restraints on development procedures have also generally been of local or, at most, regional, rather than global, importance from an oil supply perspective. There will undoubtedly be a proliferation of such requirements in the 21st century, as environmental concerns become geographically more dispersed, so encompassing developing countries and the formally centrally-planned economics. Early examples of this phenomenon can already be seen in the recognition by post-communist regimes in Russia of the need to clean-up oil production and oil transport-related pollution in its Arctic-located oilfields (Dimitrievsky, 2002); and in the World Bank's insistence on serious environmental-protection measures in connection with the exploitation of the oilfields of Chad and the export pipeline through neighbouring Cameroon. Such measures were, indeed, required as a condition for the Bank's investment in the project. Similar requirements have also been imposed on the oil export line from Azerbaijan to Turkey's Mediterranean coast. Nevertheless, the sum total of the impact and costs of such environmental protection measures still seems likely to remain of relatively minor significance at the global oil supply level.

Significant oil exploitation constraints have, however already emerged, viz. the ban in the US, including Alaska, on oil industry activities in large areas of nationally protected environments, both onshore and offshore. The US oil industry, with support from interested parties in favour of ensuring a nation self-sufficient in oil, has strenuously objected to the blanket ban over such extensive and prospective areas. That pressure did, indeed, succeed in causing the ban to be partially lifted as the US oil deficit worsened in the mid-1990s, but most of the restraints remain in place, pending Congressional acceptance of President Bush's wish to facilitate enhanced oil production in the United States (*Oil and Gas Journal*, 2003b).

Few other countries presently seem unlikely to follow suit in so restricting large areas for oil exploration/exploitation as it would take the combination of a willingness on the part of countries to forgo oil revenues and a relaxed attitude towards the issues of oil self-sufficiency and exports, to lead to the imposition of total bans on oil operations in regions with high prospectivity. Such actions would be capable of making a significant difference to the rate of reserves' discovery and, eventually, to potential flows of oil. The wealthiest of the OPEC countries – the United Arab Emirates, Kuwait, Saudi Arabia and Libya – do satisfy the conditions indicated for such an environmentally robust policy towards future oil exploitation, so the emergence of such restrictive policies on their part could be important in restraining both reserves' creation and production. Such actions, however, remain unlikely. The other more populous and less wealthy member countries of OPEC – Iran, Iraq, Venezuela, Nigeria, Algeria and Indonesia – are eminently dependent on their oil industries and, indeed, are seeking to expand them in order to meet national development objectives. They would be reluctant to have environmental considerations inhibiting these aims. Canada, with its extensive areas of arctic and sub-arctic environments is particularly exposed to possible damage from oil operations, especially with respect to the heavy oils and tar sands of Alberta – the proven reserves of which have recently been estimated at 178 billion barrels (Faithful, 2002). Should Canada decide that only environmentally friendly options can be valid for their exploitation, then there would be serious potential consequences for indigenous North American oil supply capacity in the long, as well as the short, term.

There is also a potential risk of a more general threat to offshore oil prospects. This arises from pressure by the environmental lobby – activated through the campaigning organisation, Greenpeace – to secure a ban on deepwater exploration and production because of the threat such activities are perceived to pose to the marine environment (Greenpeace, 1997). Both physical and legal action has, indeed, already been taken by Greenpeace against the companies with concessions for the exploration and exploitation of oil in the UK and Norwegian sectors of the eastern Atlantic margin. These attempts to impose an environmental restraint on deepwater oil developments were, however, successfully opposed by the UK and Norwegian governments – as well as by the oil companies – so that these specific threats to the exploitation of a potentially oil-rich region have been eliminated for the time being. The issue is, however, certain to emerge again in the near future: if not in north-west Europe, then elsewhere in the expanding world of off-shore oil activities (National Research Council of the US National Academy, 2003).

A near-future ban could, indeed, be extended to all upstream oil industry activities in the east Atlantic margin through the proactive environmental policies of the EU and the European Court (European Environment Agency, 2002). This currently seems unlikely, but it is by no means impossible, particularly if there were to be an incident which seemed to threaten serious damage to the marine environment. If restraints were imposed, either directly or through requirements that made the protection of the environment much more costly, then many billions of barrels of oil reserves would become unproducible. A European decision for such controls on the exploitation of deepwater oil reserves could then generate the possibility of similar constraints elsewhere in the world. Under such circumstances the long-term prospects for global oil supplies from deepwater offshore fields – representing the most important of the world's remaining undeveloped conventional oil frontiers (Sasanov, 2002) – would be diminished and thus change the shape of the conventional oil depletion curve shown in Figure 3.4.

To date, the limited geological knowledge of the continent of Antarctica indicates that it has oil potential but, given its adverse physical attributes, any oil wealth that may be there is not included in the estimates of the world's ultimate oil resource base (USGS, 2000). Pressures to restrict supplies from elsewhere (such as the deepwater offshore margins) could, however, lead to a desperate 'last throw' by the international oil industry to seek permission to explore the ice-covered continent, in spite of the agreement between the interested nations that it should remain an area free of potentially polluting activities. In the context of any impending oil supply crisis, the pressure for undermining that agreement would become powerful and persuasive. Likewise, any change proposed would be equally strongly fought by the environmental lobby.

Acting on the international oil industry in a more emphatic and comprehensive way is the environmental lobby's attempt to constrain the use of carbon fuels arising from its global concern for the excessive emissions of CO_2, whereby climate change, in general, and global warming, in particular, are hypothesised as the inevitable consequences (Jean-Baptiste and Ducroux, 2003). Oil, in this respect, holds an intermediate position between the more polluting coal and the significantly less polluting natural gas (Rosa, 2003). On balance, the impact of emissions controls to limit the use of carbon fuels would seem likely to be just about neutral with respect to the demand for oil, unless and until motor vehicles *en masse* are built and/or converted to run on fuels other than oil products (Douaud, A., 2002; Griffiths, 2001; Hoffman, 2001; Rifkin, 2002). The significance of such a development, given current costs, prices and policies, still appears to be at least a couple of decades in the future, but an impact can, thereafter, be expected from this fundamental change from oil to other fuels in the energy intensive transport sector (European Commission, 2003). This slowly evolving change to the use of natural gas/hydrogen for transportation fuels is, indeed, the main reason for our predicating a post-2020 declining rate of increase in oil use, compared with the recent and near future rate of increase (see Table 3.2). A more intensive pursuit of the development of clean vehicles could, however, speed up and intensify the shift to alternative vehicle fuels and thus create an earlier downward pressure on the demand for oil, particularly in the event of a broad international agreement to impose differential taxation against oil fuels (in favour of natural gas or hydrogen) (European Commission, *ibid.*) The consequential squeeze on the use of oil in the transport sector would, of course, result in the moderation of even the slow forecast growth rate in oil supply shown in Figure 3.6. Under these circumstances the predicated 2060 date for oil's peak production would be delayed, as the restraints would make unnecessary the full development of the supply side potential indicated earlier in this chapter (Groenveld, 2002).

Regional Issues in Oil Supply Prospects

Regional issues will be less important for oil supply developments than for coal (see Chapter 2 above) and natural gas (see Chapter 4 below). Although world coal resources are widely dispersed, both production and use are heavily concentrated in a small number of countries, so that prospective supply-demand relationships in the 21st century vary geographically to a marked degree. The higher costs of transporting natural gas will, likewise, continue to undermine its marketability in countries remote from producing regions.

This situation is quite different with respect to oil. Proven reserves are presently highly concentrated in the Middle East, with 65% of the

Table 3.4: Rank Ordering of the Top-10 Oil Producing and Consuming Countries in 2000

	Production			Consumption	
	m.t.o.e	% Net Exports		m.t.o.e	% Net Imports
Saudi Arabia	441	86.0	United States	898	60.5
United States	353	0	Japan	255	100
Russia	325	61.8	China	230	28.8
Iran	188	62.0	Germany	130	93.8
Venezuela	172	86.9	Russia	124	0
Mexico	171	50.9	South Korea	103	100
China	163	0	India	98	63.3
Norway	161	94.8	France	95	96.8
Iraq	129	74.1	Italy	94	95.3
Canada	127	30.7	Canada	88	0
Total of above	2230		**Total of above**	2115	
World Total	3595		**World Total**	3511	
% of Top Ten	62.0%		**% of Top Ten**	60.2%	

global total. In the absence of major surprises on the location of the world's remaining undiscovered reserves of conventional oil, it is likely that the Middle East's share of the world's ultimately recoverable conventional oil will remain above 50%. The other 50% – both proven and still to be discovered – appear likely to be found around the rest of the world in a reasonably equitable manner, except for east and south-east Asia which is poorer than elsewhere. The future exploitation of such reserves, on anything like the scale of the upstream oil industry in the Middle East, is unlikely to emerge even in the very long-term.

Meanwhile, as shown in Table 3.4, there is but a modest correlation between the rank-ordering of producing and consuming countries. Only the US, Russia, China and Canada are rank-ordered in the top-ten countries for both supply and demand. The combined output of the other six ranked producing countries is almost double the total consumption of the other six ranked consuming countries: indicating the required continuing breadth and depth for international movements of oil. The continuity of these massive international movements of oil will thus persist throughout the remaining near 30 year's growth in conventional oil use (see Figure 3.4).

Most of the movements involve inter-continental transport but, even so, there are large tonnages moving between countries in the same continent. As, for example, from Mexico and Canada to the US; from Norway to Germany and France and from Russia to many other countries of Europe. These high volumes of trade in oil reflect the ease and cheapness of transporting the commodity. In effect the costs of overcoming distance between supplying areas and markets are modest, compared with the value of oil in the market place. By contrast, inter-continental and intra-regional movements of coal are less than one-quarter the size.

The ease and low cost of oil transportation, even inter-continentally, has led to oil's pre-eminence as an international commodity, with its price determined in the most active global market-places of London, New York and Singapore. These, between them, offer 24 hours per day coverage for trading activities. This basic framework of the organisation of the international oil market seems unlikely to change for many decades, given the universal requirements for oil and the ability to move it around the world on a scale which is large enough to enable regional and national supply-demand imbalances to be equilibrated as and when they occur, even in the very short term (measured in terms of days, rather than weeks).

Nevertheless, in spite of the continuity of the international oil system as described above, there has, since the mid-1970s, been a much stronger relative rate of growth of the upstream oil industry outside the Middle East. This has resulted partly from the "away-from-Middle East oil" policies of oil importing countries in the aftermath of the oil crises of the 1970s; and partly from the inability of the state oil enterprises in the Middle East, which were created post-1973 following the nationalisation of the activities there of the multinational oil corporations, to keep their industries abreast of the new exploration and exploitation technologies which have, in the meantime, been developed elsewhere. As a result, the Americas (north and south) and Europe (excluding Russia) have become less dependent on supplies from the Middle East, the reserves of which remain hugely under-utilised (Odell, 1997a). Geopolitics, rather than the geography of oil resources, seems likely to sustain this new situation for at least the next two decades. Meanwhile, the rapidly expanding economy of south-east Asia, with inadequate oil reserves to meet its growing oil needs, has quickly achieved the status of the main oil importer from the Middle East. This will continue for at least the next 20 years and probably for very much longer (Chow, 2003; IEA, 2002c).

Thereafter, on the reasonable assumption that there will be no major changes in the existing spatial distribution of reserves of conventional oil, and that the supply of non-conventional oil will grow only slowly until 2020, there seems likely to be a period in which the world could once again become more heavily dependent on oil from the Middle East. This has, indeed, already produced an early-21st century upward pressure on prices and, of course, a major stimulus for the more rapid exploitation of non-conventional oil. The reserves of the latter are, as already shown above, more widely geographically dispersed, but with an emphasis – in terms of evaluations made to date of reserves potential – on locations in the western hemisphere, mainly Canada and Venezuela where exploitation has already begun (Meyer, 1997; Aalund, 1998) and is now entering a period of more rapid growth (*Oil and Gas Journal*, 2003b; Williams, 2003). This should enable the demands of the Americas after 2020 – when the region will still be the world's largest market for oil – to be increasingly served by oil derived from the continent's large non-conventional reserves.

The degree of Europe's increasing exposure to Middle East dependence after 2015, as production from the North Sea and associated areas ceases, at best, to grow and, at worst, to decline, will depend essentially on the success which is achieved in the more intensive and extensive exploitation of the reserves of the former Soviet Union (Considine and Kerr, 2002). The speed at which this will happen is more a question of geo-politics than of either the undoubted resource potential of these areas or the economics of their oil production and delivery systems (World Petroleum Congress, 2002a; Stinemitz, 2003). The current high levels of international oil companies' interests in the Caspian basin and in Siberia indicate their firm expectations for reserves' expansion, though there is still scepticism over the degree to which these can be realised (Kalyuzhnova, 2002). These doubts arise as a result of continuing uncertainties over the stability of the new regimes in the former Soviet republics and of their willingness to offer terms for oil exploitation (including legal certainties concerning leases and joint ventures) which justify the large investments required from the oil companies. Nevertheless, all three of the international oil industry's "mega-majors" (BP, ExxonMobil and Shell) have, since 2000, signed up to multi-billion dollar commitments with Russian oil companies for the exploitation of the country's oil (and gas) resources.

The relative paucity of the Asia/Pacific region's conventional oil reserves – only a little over 4% of the world total – has already been indicated, as has the renewal in the short-term of high growth in energy demand, including that for oil. The two major prospective areas within the region are onshore and offshore China and the South China Sea. But, China to date has generally disappointed the companies which had previously interpreted it as having high potential. More significantly, effective co-operation between China and the international oil companies, on the basis of which a much more extensive and intensive exploration effort in China could be achieved, is still in the process of being established. The continuation of this situation will at best keep oil production at a level which can only serve to limit the country's future dependence on imports (Yang Jingmin et al, 1997). In the meantime, net oil imports have grown sharply to 75 million tons compared with only 25 million tons five years earlier. Moreover, China, also became the world's second largest oil consuming country in 2002, when its use exceeded that of Japan which had hitherto been second only to the United States for more than three decades. China will almost certainly retain its new ranking on a permanent basis as its rapid economic growth continues indefinitely to stimulate energy/oil demand (Van Vuuren, 2003).

In this context, the prospects for the South China Sea become highly significant (see Figure 4.5). It helps to explain China's claim to sovereignty over virtually the whole of the maritime area, to the virtual exclusion of the claims of the other countries (Brunei, Cambodia, Indonesia, Malaysia, the Philippines and Vietnam) through their sectoral interests in the sea, based on the established median line principle. The dispute has necessarily eliminated serious interest to date in exploration, let alone exploitation. Even worse, it constitutes a continuing threat of conflict within the region in which outside powers, notably the US and Japan, could well become involved. No resolution to the dispute is yet in sight, so the potentially large oil resources of the area remain undiscovered and undeveloped (Paik and Kim, 1995). Meanwhile, China's state-owned offshore oil company (CNOOC) has intensified its own exploration activities in the country's coastal waters and has accepted bids from foreign companies for additional efforts to find more oil.

Nevertheless, the Asia/Pacific region's current dependence on more than 500 million tons of oil from the Middle East (accounting for 80% of the region's total imports and over 50% of the total demand for oil) will thus not only persist, but will also almost certainly increase over the next two decades. In the longer term, alternative large-scale imports from Russia's eastern Siberian and other Far East regions could be developed (Paik, 1995; Kennedy 2003) but, as in the case of Russian oil exports to Europe, their development will depend on the establishment of satisfactory long-term conditions for foreign investment in the relevant areas of Russia, the oil resources of which have, to date, been exploited to but a small degree.

Within the Asia Pacific region as a whole there are extensive potential reserves of non-conventional oil, but little progress has been made in defining them, let alone exploiting them. Successes elsewhere in the world in developing lower cost technologies for exploiting non-conventional oil will eventually encourage investment in the region, but no significant developments seem likely before 2020. By then, political and economic relations between the region and the Middle East may well have become close enough to ensure the acceptance of mutually advantageous interdependence between them, based on the exploitation of the Middle East's reserves pre-eminently for use in the Asia Pacific region (Odell, 1997a).

Possible New Developments

The necessity of extending the world's oil resource base to include non-conventional reserves has already been stressed. The initial hurdle in this respect has, however, recently been already overcome by the Canadian decision to declare as proven 178 billion barrels of such oil in Alberta (*Oil and Gas Journal*, 2003b). Figures 3.4 and 3.5 show how the share of non-conventional oil will rise slowly in the first two decades of the 21st century but, thereafter, will more emphatically contribute to the global supply of oil. As shown in the Figures, this contribution will have risen to over 50% when global oil production is predicated to peak in 2060; and thereafter to account for almost 90% of total supply by 2100. The anticipated medium-term progress of this new development has been described above. Its long-term implications require deeper analysis.

Beyond this critical development for expanding oil supplies, there lies an even more fundamental issue for oil's future prospects, viz. the validity of the view still overwhelmingly accepted in the West of an organic origin of oil and thus for its occurrence within quite narrowly defined areas of the earth's surface. It is from this restrictive hypothesis of the derivation of the world's oil that most estimates to date of the oil resources base have been made (but see Styrikovich, 1977; Krylov, 1997 for alternative views on oil resources). There is, moreover, a contrary view on the prospects for oil arising from the Russian-Ukrainian theory of oil's abyssal, abiotic origin. This implies that oil is not a 'fossil' fuel and may well be a renewable resource. The controversy applies also to natural gas. The highly significant implications of this alternative theory are presented in Chapter 6 of this book.

References
Aalund, L.R. (1998), "Technology and money unlocking vast Orinoco reserves," *Oil and Gas Journal*, Vol.96, October 19, pp.49–72.
Adelman, M.A. (1993), *The Economics of Petroleum Supply*, Cambridge, Mass, MIT Press.
BP (1979), *Oil Crisis… Again*, London, BP Policy Review Unit.
Campbell, C.J. (1997), *The Coming Oil Crisis*, Brentwood, Multi-Science Publishing Co.
Campbell, C.J. (2003), *The Essence of Oil and Gas Depletion*, Brentwood, Multi-Science Publishing Co.
Campbell, C.J. and Laherrère, J.H. (1998), "The end of cheap oil," *Scientific American*, March, pp.78–83.
Centre for Global Energy Studies (2001), *Oil Potential in the Middle East*, London.
Chow, L. (Ed.) (2003), "Themes in current Asian energy," *Energy Policy*, Vol.31.11.
Considine, J.I. and Kerr, W.A. (2002), *The Russian Oil Economy*, Cheltenham, Elgar Publishing.
Deffeyes, K.S. (2001), *Hubbert's Peak*, Princeton, Princeton University Press.
Dimitrevsky, A.N. (2002), "Environmental problems in developing oil and gas reserves of Russian Arctic areas." *Proceedings of the 17th World Petroleum Congress*, Vol.2, pp.551–60.
Douaud, A. (2002), "Oil gas, hydrogen and electricity: energies of the future for transport," *Proceedings of the 17th World Petroleum Congress*, Vol.1, pp.303–18.
Downey, M.W. et al. (2001), *Petroleum Provinces of the 21st Century*, Tulsa, Oklahoma, AAPG Memoir 74.
Econ Centre for Economic Analysis (1997), *Oil and Gas – a Sunset Industry?*, Oslo.
El Hinnawi, E.E. (1981), *The Environmental Impacts of Production and Use of Energy*, London, Tycooly Press.
European Commission (2003), *Towards a Hydrogen Based Economy*, Brussels.
European Environmental Agency (2002), *Energy and Environment in the European Community*, Brussels.

Faithful, T.W. (2002), "Principles to practice: striving for sustainable development in an energy megaproject," *Proceedings of the 17th World Petroleum Congress*, Vol.5, pp.77–84.

Greenpeace (1997), *Putting a Lid on Fossil Fuels*, London.

Grenon, M. (1979), *Methods and Models for Assessing Energy Resources*, Oxford, Pergamon Press.

Griffiths, J. (2001), "Fuel cells – the way ahead," *World Petroleum Congress Report*, London, ISC Ltd.

Groenveld, M.J. et al (2002), "Will the carbon age terminate before the depletion of resources?" *Proceedings of the 17th World Petroleum Congress*, Vol.1, pp.133–47

Hiller, K. (1997), "Future world oil supplies – possibilities and constraints," *Energy Exploration and Exploitation*, Vol.15.2, pp.127–36.

Hoffman, P. (2001), *Tomorrow's Energy; Hydrogen, Fuel Cells and the Prospect for a Cleaner Planet*, Cambridge, Mass, Cambridge University Press.

Hols, A. (1972), "Future energy supplies to the Free World," *EIU International Oil Symposium*, London, pp.1–24.

Holtberg, P. and Hirch, R. (2003), "Can we identify limits to worldwide energy resources?" *Oil and Gas Journal*, Vol.101.25, pp.20–6.

I.E.A. (2002c), *World Energy Outlook to 2030*, Paris, OECD.

Jean-Baptiste, P. and Ducroux, R. (2003), "Energy and climate change," *Energy Policy*, Vol.31.2, pp.155–166.

Kalyuzhnova, Y. et al (Eds.) (2002), *Energy in the Caspian Region*, Basingstoke, Palgrave.

Khartukov, E.M. (1997), "The control of Russia's oil," *Energy Exploration and Exploitation*, Vol.15.2, pp.117–26.

Krylov, N.A. et al (1997), "Exploration concepts for the next century", *Proceedings of the 15th World Petroleum Congress*, Beijing.

Kennedy, C. (2003), "The expansion of Russia's Siberian export capacity," *Oxford Energy Forum*, August, pp.11–12.

Laherrère, J.H. (1997), "Production, decline and peak reveal true reserves figures," *World Oil*, pp.77–83.

Laherrère, J.H. (2003), "Future of oil supplies," *Energy Exploration and Exploitation*, Vol.21.3, pp.227–267.

Lainier, D. (1998), *Heavy Oil: a Major Energy Source for the 21st Century*, Edmonton, UNITAR.

Lynch, M.C. (1997), *The Wolf at the Door or Crying wolf: Fears about the next Oil Crisis*, Cambridge, Mass, Centre for International Studies, M.I.T.

Martinez, A.R. and McMichael, C.L. (1997), "Classification of petroleum reserves," *Proceedings of the 15th World Petroleum Congress*, Beijing.

McCabe, P.J. (1998), "Energy resources: cornucopia or empty barrel," *AAPG Bulletin*, Vol.82.11, pp.110–134.

Meyer, R.F. (1997), "World heavy crude resources," *Proceedings of the 15th World Petroleum Congress*, Beijing.

Meyer, R.F. and Olson, J.C. (Eds.) (1981), *Long-Term Energy Resources*, Barton, Pitman.

National Energy Board of Canada (2000), *Canada's Oil Sands, Supply and Market Outlook*, Calgary.

National Research Council (2003), *Oil in the Sea III: Input and Effects*, Washington, DC, National Academies Press.

Odell, P.R. (1964), *The Economic Geography of Oil*, London, Bell.

Odell, P.R. (1973), "The future of oil: a rejoinder," *Geographical Journal*, Vol.139.9, pp.436–454.

Odell, P.R. (1979), "World Energy in the 1980s: the significance of non-OPEC supplies," *Scottish Journal of Political Economy*, Vol.26.5, pp.215–255.

Odell, P.R. (1981), "Prospects for and problems of the development of oil and gas in developing countries," *National Resources Forum*, Vol.5.4, pp.317–26.

Odell, P.R. (1985), "East-West differ on estimates of oil reserves," *Petroleum Economist*, Vol.52, pp.329–31.

Odell, P.R. (1994b), "World resources, reserves and production," *Energy Journal Special Issue*, pp.89–114.

Odell, P.R. (1997a), "The global oil industry: the location of oil production," *Regional Studies*, Vol.31.3, pp.309–20.

Odell, P.R. (1998), "Oil and gas reserves: retrospect and prospect," *Energy Exploration and Exploitation*, Vol.16.2, pp.117–124.

Odell, P.R. (2001), *Oil and Gas: Crises and Controversies, 1961–2000*, Vol.1, *Global Issues*, Brentwood, Multi-Science Publishing.

Odell, P.R. and Rosing, K.E. (1980, 1st Edition and 1983 2nd Edition), *The Future of Oil; a Simulation Study of the Inter-Relationships of Resources, Reserves and Use, 1980–2080*, London, Kogan Page.

Oil and Gas Journal (2003b), "Future Energy Supply," *Oil and Gas Journal*, Vol.101, Nos.27–32.

Paik, K-W. and Kim, D-K. (1995), "The Spratly islands' dispute with China," *Geopolitics of Energy*, Vol.17.10, pp.5–10.

Rifkin, J. (2002), *The Hydrogen Economy*, London, Penguin Books.

Sasanov, S. (2002), "The deep-water challenge," *World Petroleum Congress Report*, London, ISC Ltd, pp.120–7.

Shell, IPC (1995), *Energy in Profile*, London, Shell Briefing Service.
Shell IPC (1998a), *A Commitment to Sustainable Development: the Global Scenarios Project*, London.
Shell IPC (2001), *Energy Needs: Choices and Possibilities: Scenarios to 2050*, London.
Smith, N. and Robinson, G.H. (1997), "Technology pushes reserves crunch date back in time," *Oil and Gas Journal*, Vol.95, April 7, pp.43–50.
Stinemetz (2003), "Russian oil sector rebounds," *Oil and Gas Journal*, Vol.101.22, pp.20–30.
Styrikovich, M.A. (1977), "The long-range energy perspective," *Natural Resources Forum*, Vol.1, No.3, pp.252–63.
United States Geological Survey (2000), *World Petroleum Assessment*, Reston, Virginia, Government Printing Office.
Van Vuuren, D. et al. (2003), "Energy and emissions scenarios for China in the 21st century", *Energy Policy*, Vol.31, No.4, pp.369–88.
Verbicky, E. (1998), "Oilsands: a growing and viable alternative to conventional oil." *Petroleum Economist*, Vol.65, No.1, pp.21–3.
Warman, H.R. (1972), "The future of oil," *Geographical Journal*, Vol.138.3, pp.287–97.
Watkins, G.C. (2000), *Characteristics of North Sea Oil Reserves Appreciation*, Center for Energy Policy Research, Massachusetts Institute of Technology (MIT-IEEPR 2000-08 WP).
Williams, B (2003), "Heavy Hydrocarbons to play key role in future energy supply," *Oil and Gas Journal*, Vol.101.29, pp.20–7.
Williamson, H.F. (1963), *The American Petroleum Industry*, 1859–1959, Evanston, North Western University Press.
World Petroleum Congress (2002a), "New hydrocarbons provinces of the 21st century," *Proceedings*, Vol.2, Forum 2, pp.87–176.
Yang Jingmin et al. (1997), "Analysis of world oil supply and demand and the development trend of the Chinese petroleum industry," *Proceedings of the 15th World Petroleum Congress*, Beijing.

Chapter 4: Natural Gas – The Prime Energy Source for the 21st Century

Resource Abundance

After a number of false dawns, from the mid-1970s to the early 1990s, for an anticipated near-future major expansion of natural gas for both geo-political and environmental reasons (viz. diversification away from energy dependence on Middle East oil and much reduced CO_2 emissions compared with other carbon fuels, respectively), global gas production finally seems to have entered a period of continuing and significant expansion. Indeed, in the last decade of the 20th century world gas production – and consumption – grew almost 50% more quickly than that of energy overall; albeit at an average annual rate of growth of only 2.1%. This development seems at last to negate the earlier well established and widely-held views that the natural gas resource base and/or the energy markets which gas could serve were too limited to make possible its emergence as a third significant energy source alongside coal and oil (Marcetti, 1978). Nevertheless, by 2000 gas contributed almost 24% of global energy use; less than 2½% behind the contribution of coal. Expectations for its continuing expansion in both absolute and relative terms are now widely accepted (Odell, 1998a; Shell, 2001, World Petroleum Congress, 2002a; Adelman and Lynch, 2003).

Along with the expansion of demand, there was an even more rapid growth in proven reserves, from 57Gtoe in 1975 to almost 135Gtoe by 2000. After taking into account the production of about 40Gtoe over the same period, this implies a more than tripling of discovered reserves over the 25 years. The reserves-to-production ratio (based on current annual production of about 2.2Gtoe) increased to over 60 years. Gas production has expanded in all major regions except the former Soviet Union, as have the remaining proven reserves of all the regions except North America. In brief, all major indicators point to gas expansion as the norm, with a firm expectation that the process will continue (Thackeray, 1998; International Gas Union, 2003).

Concern about the future availability of gas at the global level has never become an issue in any meaningful way (Delahaye and Grenon, 1983) in marked contrast to the many previously perceived concerns for the adequacy of the world's oil resources (see Chapter 3). There have, however, been recent fears for the continuity of supply availabilities in three regions, viz. the US, Western Europe and the former Soviet Union.

In the US, these fears had some justification, given the maturity of the industry (which dates back to the early years of the 20th century) and the low reserves-to-production ratio (only a little over 10 years) with which the industry has worked for over 15 years. Recent reappraisals of old gas-producing areas, together with new, mainly offshore, reserves discoveries, have, however, converted the earlier pessimism into moderate optimism for continued growth in both reserves and annual production (World Petroleum Congress Proceedings 2002a, Vol.4). The level of production is, indeed, now getting back closer to the earlier all-time high, over 30 years ago in 1972. Nevertheless, the country now also needs to import about 115 Bcm (103mtoe) in order to meet burgeoning demand – in spite of a doubling of the gas price since 1990. The implications of this for the future of the US natural gas industry in the 21st century are discussed later in this chapter.

In Western Europe recurring fears of gas scarcity were taken so seriously in the mid-to-late 1970s – at a time when indigenous production was in its infancy – that restraints on gas use for power generation were directed by the European Commission. At the same time, the Netherlands, then the principal European gas producing country, prohibited any additional exports. Later, the UK and Norway also deliberately constrained production. The fears were, however, entirely irrational, given that they were based on incorrect assumptions, viz. first, that the gas supply was price inelastic and, second, that indigenous proven reserves, based on a very limited exploitation of the potentially gas-rich provinces of north-west Europe, told the whole story of future supply possibilities. In reality, it was the limitations on demand and inappropriate government policies that jointly inhibited gas exploration and exploitation as they made investments in upstream developments uninteresting and unrewarding (Odell, 1988a). The situation and outlook have, however, been reversed since 1990 with a resulting 60% increase in West European production. There does, nevertheless, still remain a tendency for the countries concerned (except Norway) to report reserves and prospects conservatively, so persuading energy policy makers that high gas dependence is unwise (Odell, 1995 and 2002).

One notable element in the fundamentally-changed politico-economic situation, following the break up of the USSR, has been the maintenance of the level of declarations of proven reserves in the former Soviet Union; and especially in Russia, whose reserves constitute 86% of the FSU's total. This reserves position has been achieved in spite of the political and economic traumas that affected the newly independent countries of the former Soviet Union, and in the context of a declining demand there for natural gas (from 666 billion cubic metres in 1991 to a low of 465 billion cubic metres in 1999) as a result of those problems. The reserves-to-production ratio for the FSU as a whole is now over 80 years, well ahead of the world average of 61 years (Vyakhirev et al, 1997)

Given these outlooks in 2000 for natural gas in the three most important world markets for its production and consumption then, in an evaluation of its long-term future, natural gas can be seen to be starting its 21st century role from a much more favourable base than that for oil. Proven global reserves simply as declared – but without taking account of the inevitable appreciation which will emerge from the continuing development of the industry – could keep global gas production growing at about 3% a year for some 25 years. Even then about half of currently proven reserves would still remain unused in 2025. Thereafter, of course, the continuing ability of conventional gas production to grow depends on additional reserves having been found in the meantime – but this is already a certain prospect (Cornot-Gandolphe, 1995 and USGS, 2000).

Indeed, large volumes of additional reserves are widely and generally expected because most of the existing gas producing provinces have been developed relatively recently and thus remain areas for continuing investments in additional relatively low cost exploration and production. There are also other large areas of potential, both onshore and offshore, which remain entirely or almost entirely unexplored. The current range of estimates of recoverable reserves is shown in Table 4.1. This shows that 200Gtoe of natural gas had been used or was recognised as proven reserves by 2000; and that another 198–303Gtoe of additional reserves are also expected. These are described as being 'extremely conservative' assessments for some regions, so that 'additional reserves could exceed the total shown in the table' (International Gas Union, 2003). Indeed, the United States Geological Survey in its recent comprehensive world-wide evaluation of remaining undiscovered conventional gas resources arrived at a mean value of almost 400Gtoe (USGS, 2000).

In plotting a global production curve for depleting the volume of natural gas indicated (see Figure 4.1), two assumptions have thus been made:

> *first*, that the supply of gas will increase at a rate which ensures that the combined growth rate in the use of hydrocarbons (oil and gas) can be met. This will require a growth rate in gas supply in excess of 2% per annum for the first half of the century and thereafter at a progressively lower rate of growth until 2090.

> *second*, that the growth curve for conventional natural gas will persist until the time at which the use of about 40% of the mid-point of the range of the ultimately recoverable resources is approaching, viz. by the early 2040s. Thereafter, the slope of the curve will gradually fall away under the pressure of an increasing reserves restraint, until the peak production of conventional gas is

Table 4.1: World conventional gas reserves and resources, by region and percentage depletion by 2000

Region	Gtoe				Percentage Depletion of Reserves by 2000
	Produced to 2000	Proven reserves	Estimates of additional reserves	Ultimately recoverable reserves	
North America	27.3	6.7	30–52	64–86	31.7 to 42.7
Central and South America	3.1	6.2	7–22	16–31	10.4 to 19.4
Europe (excluding FSU)	7.5	4.7	5–14	17–26	28.8 to 44.1
Former Soviet Union	16.7	51.0	96–110	164–178	9.4 to 10.2
Middle East	4.1	47.3	29–50	80–101	4.1 to 5.1
Africa	2.2	10.1	5–14	17–26	8.5 to 12.9
Asia Pacific (excluding FSU)	3.8	9.3	26–41	39–54	7.0 to 9.7
Total	**64.7**	**135.2**	**198–303**	**397–503**	**12.9 to 16.6**

Source: BP Statistical Review of World Energy, 2001; IIASA, Global Energy Perspectives to 2050 and Beyond, 1995; Rogner, H-H., An Assessment of World Hydrocarbon Resources, 1996; International Gas Union, 22nd World Gas Conference, Tokyo, 2003.

reached in 2050. By then about 220Gtoe of the world's ultimately recoverable conventional gas of 450Gtoe (the mid-point of the range shown in Table 4.1) will have been used. As shown in Figure 4.1, the decline curve then sets in, and by 2100 conventional gas supply is down to less than half of its peak rate in 2050. By that date the ultimate resource base as presently defined will be almost 90% depleted.

Given the results of these assessments of the future supply prospects for conventional natural gas, it is self-evident that the maintenance of a rising availability of hydrocarbons each year through the second half of the 21st century will depend on the exploitation of non-conventional gas resources (Delahaye and Grenon, 1983). The potential availability of these even in the recent past has been only modestly evaluated – so modestly, indeed, that they did not figure at all in the International Gas Union's 1997 presentation of world gas prospects (IGU, 1997), except as an unspecified component in the gas reserves data for North America (and thus included as an unknown element in the additional gas reserves' figure for that region in Table 4.1). For most of the rest of the world, the potential for gas recovery from the range of unconventional habitats (viz. coal-bed methane, tight formation gas, gas from shales and gas remaining *in situ* after conventional production) has not yet become a relevant question as a result of the large remaining conventional gas resources in relation to demand expectations over the next 40 to 50 years.

Thus, only speculative figures exist for the global non-conventional gas resources base. One such set of figures has been derived from the in-depth study made by the Vienna-based International Institute for Applied Systems Analysis in the mid-1990s, *Global Energy Perspectives to 2050 and Beyond* (IIASA, 1996). This indicates potentially recoverable resources in the range of 779–948Gtoe – of which 138 are already known and considered to be technically recoverable. As shown in Table 4.2, non-conventional gas resources are geographically distributed across all the world's continents (except Antarctica which was not included in the analysis). The study also made estimates of possible resources of gas in hydrates, with the global and regional results as also shown in Table 4.2. These figures were gigantic (compared with other non-conventional gas resources, let alone set against conventional gas' prospects), but have since been challenged (USGS, 2001; Cherkashov, G.A. and Solowiev, V.A, 2002). Further attention is given to these prospective resources' exploitation later in this chapter.

Table 4.2: World non-conventional gas resources, by region

Region	Coal-bed methane, tight formation gas and gas from shales and gas remaining after conventional production	Gas hydrates* and geopressured gas
	In Gtoe	
North America	210–230	6089
Central and South America	87–95	4567
Europe (excluding FSU)	32–40	761
Former Soviet Union	139–181	4186
Middle East	86–112	190
Africa	27–32	381
Asia Pacific (excluding FSU)	198–258	2474
Total	**779–948**	**18647**

* A more recent evaluation (Cherkashov and Solowiev, 2002) assesses the world resources of potentially recoverable hydrates at only 8–10% of the estimated total in this column: but it does not give a regional breakdown, see text.

Source: H-H. Rogner, *An Assessment of World Hydrocarbon Resources*, IIASA 1996

Production Potential Overall

For the purposes of modelling the 21st century supply curve, production of non-conventional gas is predicated to begin at the time (around 2020) when conventional gas production may require some supplementation so that an availability of gas sufficient to sustain an annual growth of ±2% in the overall hydrocarbon supply can be achieved. Note that this is a pessimistic view of the timing of the initial use of non-conventional gas, given that small – but nevertheless significant – volumes of such gas are already being recovered in the US, mainly from coal measures (*Oil and Gas Journal*, 2003a). The amounts which will be produced in the short to medium term, however, seem likely to be modest, because of competition from lower cost conventional gas production, except in the special circumstance of the United States where over 26% of current global gas use is concentrated. In the global context, it is likely to be the second quarter of the 21st century before the quantity of non-conventional gas is large enough to impact significantly on the shape of the production curves in Figure 4.1. It is, nevertheless, reasonable to assume that once the technology of producing gas from coal measures or from other non-conventional habitats in the US has been perfected, it will rapidly spread to other areas of significant potential (USGS, 1997; *Gas Matters*, 2003). This development will lead thereafter to a rapid build-up of production, providing that demand for natural gas continues to grow (as we forecast), and that energy prices remain high enough to justify the investment required (as we also predicate). This process of increasingly intensive and extensive non-conventional gas exploitation will, we suggest, begin in earnest in the 2030s (although it will have begun earlier in the US itself).

As shown in Figure 4.2, non-conventional gas will from the 2020s to 2050 complement output from the decelerating rate of increase in conventional gas supplies. After 2050, when conventional gas supplies start to decline, defined non-conventional reserves and their potential production will have become substantial enough to enable the global production of gas to continue to increase. Thus, it is predicated, non-conventional gas will take over as the more important component in the total supply in the mid-2060s. On the assumption that the ultimately recoverable reserves of non-conventional gas total 650Gtoe (equal to 80% of the low end of the range of the IIASA defined resource base of 780Gtoe – as shown in Table 4.2 – and so excluding any gas recovery from geo-pressured gas and gas hydrates), then output will reach an 'inevitable' peak towards the end of the 21st century (see Figure 4.1) when its cumulative production reaches about 340Gtoe; that is 50% of the defined level of ultimately recoverable reserves.

Meanwhile, as shown in Figure 4.2, the peak of total conventional plus non-conventional gas output will also occur in 2090 with a cumulative production from 2001–90 of about 650Gtoe. More than fifty years prior to that, however, in the 2030s, the annual contribution of gas to the global energy economy will, as shown in Figure 1.5, have exceeded that of oil: while by 2064 cumulative gas use in the 21st century will exceed that of oil (see Fig. 3.7). Of greater importance, however, is the continuing ability until 2080 of natural gas to continue to sustain the annual increase in the hydrocarbon supply required by the by-then diminishing rate of growth in the demand for energy: as a result of the factors considered in Chapter 1. There is thus no really significant shortfall in hydrocarbons' availability to satisfy the overall demand for energy, except in the last decade of the century. This exception is caused by the post-2090 decline from the peak production of natural gas (see Fig. 4.2), whereby there is a major boost to the late-21st century requirement for renewables. Cumulative renewals production in the 1990s is 20% higher than in the previous decade (see Fig. 1.5).

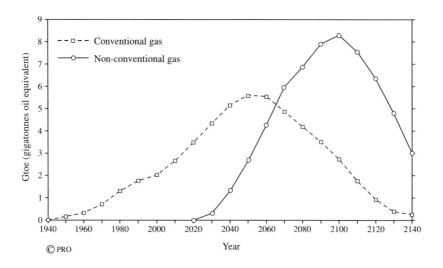

Figure 4.1 Production curves for conventional and non-conventional gas, 1940–2140

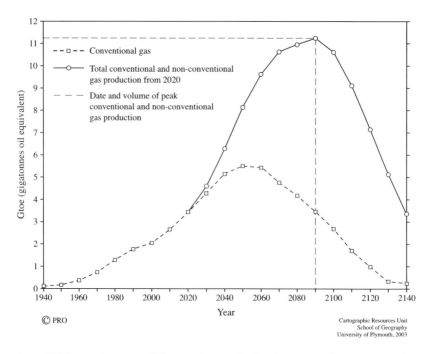

Figure 4.2 The complementary relationship of conventional and non-conventional gas production, 1940–2140

This development in the last decade of the 21st century apart, we thus conclude that there is not a significant amount of energy which will have to be provided from non-carbon sources in the 21st century. Indeed, the requirement will be a relatively modest one in the context of the total carbon energy contribution to world energy supply. Total global energy demand over the 100 years will be 2360Gtoe. This will comprise 520Gtoe of oil, 695Gtoe of natural gas, an environmentally-restrained contribution of coal of 460Gtoe: and an arithmetically-derived requirement for renewables over the century of some 685Gtoe. The latter constitutes 29% of the total amount of energy supplied and demands an annual average growth rate in use over the century of just over 3% per annum.

But this prospect for renewables, in the context of neither significant subsidies for the development of renewable energy, nor the imposition of significant environmental taxes on the use of carbon energies, could be reduced by almost a half were the supply of oil and gas only 28% higher than predicated, viz. another 350Gtoe. The same result could also be achieved by the exploitation of about 10% more of the world's coal resources which are already known to exist. Indeed, coal resources could substitute all the renewables without any physical supply or cost/price problem. Just as prospective as either of these options is the likelihood that additional gas supplies could well emerge in the second half of the century from the initial period of exploitation of gas hydrates (Lowrie and Max, 1999). Such a new source of supply would barely make a tiny dent in the speculative 18,647Gtoe of such resources indicated in Table 4.2. Even the recent 90% lower estimates of 1600 to 2000Gtoe of exploitable gas hydrates reserves (Cherkashov, and Solowiev, 2002) would be only modestly depleted by a late 21st century demand for up to 350Gtoe of such gas as a supplement to the gas produced from conventional and other non-conventional habitats.

Another 50 years or more of continuing scientific advances and developments in engineering capabilities for gas hydrates production (Carroll, 2003) would seem to give more than enough time to enable a small number of the 70 regions worldwide with gas hydrates on the sea bottom (Cherkashov and Solawiev, 2002) to be brought to commercial exploitation – as a means whereby the late 21st century supply-side gap of carbon fuels revealed by the foregoing analysis could be filled. Equally capable, however, of undermining the need for renewables is the prospect of a world that moves to higher levels of efficiency in energy use (Bradley, 2003) and in so doing creates an opportunity for stringent constraints on the overall demand for energy, so that carbon fuels could secure all but a small percentage of the total market. If this development were to be accompanied by policies and technologies which secure the sequestration of CO_2 emissions, then renewables would become even further disadvantaged by their high costs.(IEA, 2003b.)

Natural Gas' Environmentally Friendly Characteristics

Unlike oil, the production and transportation of gas has not led to environmental problems in terms of adverse effects on landscapes or marine conditions. Explosions and accompanying fires do, of course, constitute a danger to life, but they have not occurred frequently. While increasing the production and use of conventional gas to roughly three times its present level will exacerbate the problems noted above, they hardly seem likely to constrain pipelined supplies, except on a temporary basis and in particular locations. More concern must, however, arise from the now rapidly growing bulk international movements of gas in its liquefied state (LNG) by ocean going tankers. Such movements and the accompanying liquefaction/loading and unloading/regasification facilities are inevitably more hazardous and will need to be closely monitored and regulated as volumes expand. Current annual movements are under 0.15Gtoe (compared with 0.4Gtoe of pipelined gas), but this is expected to increase threefold by 2025 and by the end of the century it could account for up to 10% of the expected annual deliveries of almost 11Gtoe of natural gas to global markets (Quinn, 2000).

In spite of the world's increasing use of carbon energy in the 21st century (as specified in this and the two preceding chapters), the massive substitution of coal and oil by natural gas will restrain the rate of growth in anthropogenic-created emissions of CO_2 by an estimated ±15%, compared with the emissions which would have occurred had the year 2000 percentage contributions of the three carbon fuels to the total supply of energy remained unchanged. This percentage reduction in CO_2 emissions from the changing carbon fuel mix will, moreover, be further enhanced by the expected increasingly successful processes of sequestering CO_2, rather than allowing it to go to the atmosphere. (Torp, 2001; Freund, 2002; Moritis, 2003). As a result, CO_2 emissions overall from carbon energy use in 2100 seem unlikely to be much more than twice their 2000 level.

The lower CO_2 emissions from gas per unit of energy used, compared with coal and oil, make gas the preferred carbon fuel (Energy Information Administration, 1994a; Freund, 2002; Gregory, 1998; Jean-Baptiste, 2003; World Petroleum Congress, 2002c). Thus, many industrialised countries have already specified a much increased gas use as the main means by which they can meet their lower emissions targets under the terms of the Kyoto Protocol. Many developing countries also see increased gas use as a principal means not only of cleaning up their cities, but also achieving more generous developmental assistance in return for lower CO_2 emissions (Paik, 2002; Bartsch and Müller, 2000; IEA, 2002b; Chow, 2003). The 21st century implications for this steadily rising preference for natural gas – even over oil – are shown in Figure 4.3. Gas supply overtakes that of oil in the 2040s and in the last decade of the century it accounts for about 73% of the world's hydrocarbons supply.

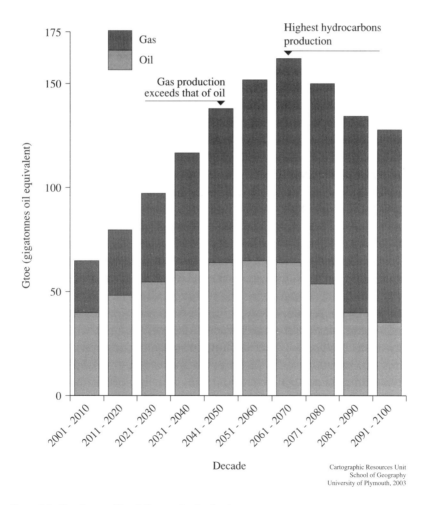

Figure 4.3 21st Century Oil and Gas supplies by decade

Methane itself is, however, also a greenhouse gas which, when released to the atmosphere, is over ten times more significant than CO_2 on a unit basis, so causing concern over its contribution to global warming and rising sea levels. More than insignificant volumes of gas can be lost to the atmosphere in production operations, while pipeline leakages are a significant problem in some countries, notably those of the former Soviet Union (Van de Vate, 1997). More effective engineering and improved operational systems will become increasingly required as volumes of gas supply increase more than five times over the century, particularly as much of this increase will be derived from the production and transport of gas in countries without much previous experience of the industry; for example, in Latin America and the Asia Pacific region. A failure to achieve high enough standards in preventing methane emissions could, in due course, produce an institutional constraint on gas production, the impact of which might well become great enough to moderate the upward slope of the supply curve shown in Figure 4.2. Only in extreme circumstances, however, is it possible to envisage as serious an environmental constraint on gas production and use as that faced by coal – and oil.

Regional Gas Markets
Viewed regionally, natural gas supply developments are subject to greater uncertainty than in the case of oil. Oil is so easily and cheaply transported that delivery restraints – even for countries remote from areas of production – are of relatively minor significance. Oil, in essence, is available virtually everywhere on demand (except in the very short term). Natural gas developments, on the other hand, have been and remain exposed to regional considerations. In fact, geography can be designated as the principal influence at work in determining the speed and character of emerging global gas supply and demand patterns. Such dominant regional variations make the global presentation of the supply prospects for gas somewhat less robust than those for coal and oil, in that decisions on gas developments within individual regions are capable of creating significant changes to the timing of the evolution of global supply. This has happened in recent decades as, for example, from decisions in the mid-1970s by the European Commission and individual European governments to restrict the growth of gas demand and hence of gas supply; and in the United States in the last decade of the 20th century as a result of inappropriate market liberalisation measures which also

constrained supply, so leading to much higher prices. These, in turn, constrained demand.

Since such factors emerge as a result of a wide range of contributory factors which are internal to the region and of varying importance over time, it is impossible in a short overview to present specific conclusions on future regional supply issues with a great deal of confidence – even for the short to medium term, let alone for the whole of the 21st century. The following comments thus represent nothing more than indicators for likely developments of significance in each of the seven world regions, as designated by the 20th International Gas Union in 1997 and now used by that organisation as the basis for its regional reports (see also Lerche, 2002).

North America

Virtually one-third of world gas use in 2000 was in North America, with the United States accounting for 85% of the regional total (see Table 4.3). As the US has long been squeezed for indigenous gas reserves – initially because of stringent regulatory controls on production and transportation over almost two decades to the mid-1990s and recently by an inappropriate form of market liberalisation – its gas industry's production stagnated at 500 to 550 Bcm per year.

Deregulation and liberalisation has thus had to be supplemented by Federal government support designed to secure the re-expansion of proven reserves to a degree sufficient to offset depletion, in the context of an annual production of gas which became inadequate to meet an

Table 4.3: Contribution of natural gas to total energy and carbon fuel use by region, 2000

Region	Gas use (mtoe)	Share of total energy use (%)	Share of carbon fuel use (%)
North America	683.6	25.3	29.4
Central and South America	83.7	18.5	25.8
Europe (excluding FSU)	412.9	22.0	27.2
Former Soviet Union	492.3	52.5	58.7
Middle East	173.4	44.4	44.5
Africa	50.0	18.0	19.5
Asia Pacific (excluding FSU)	261.6	10.7	12.0
World total/average	**2,157.5**	**23.7**	**27.2**

Source: *BP Statistical Review of World Energy*, 2001

annually increasing demand; from 540 Bcm in 1990 to 655 in 2000. Though the increased demand has stimulated technological advances in exploration and production methods with positive results, particularly in the deep waters of the Gulf of Mexico (OGJ, 2003), the additional volumes of gas production achieved to date have done little more than offset declines in output from pre-existing producing regions, including the Gulf's shallow water fields. Meanwhile significant growth in non-conventional gas extraction from tight sandstones, gas shales and coalbed methane in the mid-west and the Rocky Mountains has come to contribute almost 10% of total production (*Oil and Gas Journal*, 2003a).

Thus, the US market has attracted a large inflow of supplies both from newly developed and from more intensively exploited Canadian reserves (Canadian Gas Potential Committee, 2001). These reserves, plus almost equally potentially important, but still largely undeveloped, reserves in Mexico, also offer hope for North America's short and medium-term future supplies of conventional gas. They will, nevertheless, have to be complemented by large liquefied natural gas (LNG) imports from South America and other parts of the Atlantic basin (such as Nigeria and other west African countries) in the medium to longer term. Indeed, foreign LNG's contribution to US gas supply is likely to increase from its present less than 1% to over 20% by the early 2020s (Jensen, 2003). This becomes possible in the context of the continuation of the higher gas prices to which the US has succumbed since the turn of the century – after decades of gas prices well below the oil equivalent price.

Meanwhile, as demonstrated above, the United States leads the world in the commercialisation of non-conventional gas. Its reserves of these, on the basis of present knowledge, appear to comprise about 25% of those assessed for the world as a whole (see Table 4.2). Their accelerated exploitation, already well under way in respect of coal measure methane production, could if need be, in the context of today's higher prices, probably sustain an expansion of North America's natural gas production at the average 2% per annum expected increase in demand for most of the 21st century (Oil and Gas Journal, 2003a).

This early significant North American dependence on non-conventional gas supplies seems likely to remain unique to the region for at least the first quarter of the 21st century, reflecting the generally much more generous conventional gas reserves' availability – relative to lower gas use – in the rest of the world.

Central and South America
This region is second bottom to Africa in its current use of gas, and is below the global average in terms of the contribution of gas to energy use. Over the past decade, however, the region's natural gas use has grown twice as fast as energy use overall. Its proven reserves are in excess of 77 years of present production and it is already indicated as having twice as many additional reserves of conventional gas (see Table 4.1). Intensive exploration of its proven and additional reserves had until the last decade of the 20th century been constrained by institutional factors and, equally fundamentally, by the generally long distances between the potentially available reserves and the centres of energy use (Odell, 1981 and 1984).

Institutional changes in most countries of the region have now not only freed-up markets, but have also encouraged the flow of capital to the gas industry – in large part at the expense of additional long-distance power transmission lines from remote and high-cost hydro-electricity projects. These changes in energy strategy suggest a fast and radical evolution of a geographically more intensive and extensive gas industry (IEA, 2003a). Much infrastructure is under construction for gas transmission and distribution so that market growth will be rapid in the short-term; with a likely more than doubling of production and use by 2010. By that date there will be at least a skeletal network of gas facilities inter-connecting Argentina, Bolivia, Brazil, Chile and other neighbouring countries in the southern part of South America. A second integrating system is also emerging –albeit more slowly in the face of adverse political conditions – in the Andean region from Venezuela to Peru. A link between the two systems could, however, be developed by 2020 (Kurtz, 1997).

In the longer-term, gas from non-conventional habitats – especially from shales and tight formations – offer large potential resources, but they all await exploitation pending the expansion of demand. The continent is also considered to be especially rich in offshore gas hydrates (see Table 4.3); but, as emphasised above, no progress can be expected in the recovery of such resources until the technology for gas hydrates' exploitation has been proven elsewhere in the world, where much larger gas demands justify the research and development expenditures which will be required to achieve a breakthrough. This is unlikely until well into the second half of the 21st century.

Europe (excluding the FSU)
Europe's large natural gas industry is a relatively recent phenomenon dating only from the discovery of the giant Groningen field in the

Netherlands in 1959 (Odell, 1969). This discovery not only initiated the subsequent intensive search for hydrocarbons in the North Sea and adjacent areas, but also enabled a gas transmission system to be built in one of the most heavily industrialised and most densely populated regions of the world. In spite of serious misinterpretations of both the supply and demand potential in the late 1970s and the 1980s which led to over a decade of near-stagnation in indigenous gas supply (Odell, 1988a and 1992), the industry expanded by over one-third in the early 1990s, with both more intensive and geographically more extensive developments. Europe is now third only to North America and the FSU in the use of gas (see Table 4.3) and, likewise, the world's third largest gas producing region. The contribution of gas to the European energy economy exceeded 23% in 2000 and is still rising, as gas substitutes both coal and oil, for a combination of economic and environmental reasons. A further near doubling of the industry's size by 2020 is highly likely, for which most of the gas required will come from Europe's own increasing gas reserves (Odell, 1995). Of the conservatively estimated ultimately recoverable reserves of 17–26Gtoe of conventional gas (see Table 4.1), only about 9Gtoe have been used to date (USGS, 2000). Most of the unexploited reserves lie under the very extensive Norwegian continental shelf – stretching from the North Sea to the North Cape and into the Barents Sea, divided between Norway and Russia. These reserves will, if fully exploited, be able to supply much of Europe's demand for gas for at least the first quarter of the 21st century (Norwegian Ministry of Petroleum and Energy, 2002).

Nevertheless, given the anticipated large-scale growth in demand indigenous reserves will have to be increasingly supplemented by some of the readily available and relatively low cost resources of Russia and Algeria – and possibly from a range of other countries as well (for example, Turkmenistan, Iran and Libya by pipeline and from elsewhere as LNG). These external sources are not only abundant, but are also relatively low-cost, compared with Europe's more expensive remaining off-shore reserves (Mabro and Wybrew-Bond, 1999). Their impact on keeping prices under control will encourage the further expansion of the European gas market. As an average 2% per annum demand growth can be expected over the next 20 years. Europe will become an increasingly attractive region for supplies from additional external sources over this period.

Its gas future in the medium term – and even more so in the longer term – will depend on the further interconnection and integration of the external supplying countries with the European market, so effectively creating a continental gas system which will eventually extend from the Atlantic Ocean and the North Sea/Barents Sea in the west to the Caspian Sea in the east, and from North Africa to the Urals (see Figure 4.4). A still major undetermined issue is whether the gas-rich countries of the Middle East will be able to compete in the market for pipe-line gas to Europe with their own gas export potential (Estrada, 1995; Mabro and Wybrew-Bond, 1999). If so, then Europe will have access to sufficient conventional gas resources to meet increased demand for the whole of the 21st century. In this advantageous context, there is little likelihood that Europe will become other than modestly interested in LNG brought by tankers from distant exporting countries, let alone in the exploitation of non-conventional gas resources which (except in locally favourable circumstances) appear to be late 21st century options at best.

The Former Soviet Union
Apart from Russia, five other former Soviet republics have large gas reserves and considerably more potential for enhancing those reserves. Collectively, they constitute the part of the world with the largest conventional gas reserves. As shown in Table 4.1, estimates of proven remaining reserves exceed 50Gtoe (almost 35% of the world's total), while assessments of likely economically producible additional reserves range from 96Gtoe to 110Gtoe (around 36% of the world total). The collective annual use of the countries concerned is currently just over 0.5Gtoe, so that even the remaining proven reserves give a reserves-to-production ratio of about 100 years. When the anticipated additional reserves, amounting to twice as much more gas, are taken into account, there is clearly a powerful motivation for long-term efforts by Russia, Azerbaijan, Kazakhstan, Turkmenistan, Ukraine and Uzbekistan to achieve more extensive and intensive exploitation of their natural gas, related not only to ensuring the continuity of their 51% level of gas dependence in their own energy economies, but also to enable them to exploit further developable market opportunities in Europe (Stern, 1995) and prospectively in China, India and other countries of east and south-east Asia. (Wybrew-Bond and Stern, 2002; Stinemetz, 2003).

The existing pipeline links from Russia to Europe and projects and proposals for additional links (see Figure 4.4) already reflect this situation (Bakhtiari, 2003), so that the development of a gas system fully integrated with the European demand area is a near certain mid-21st century prospect: both directly from Russia, and indirectly

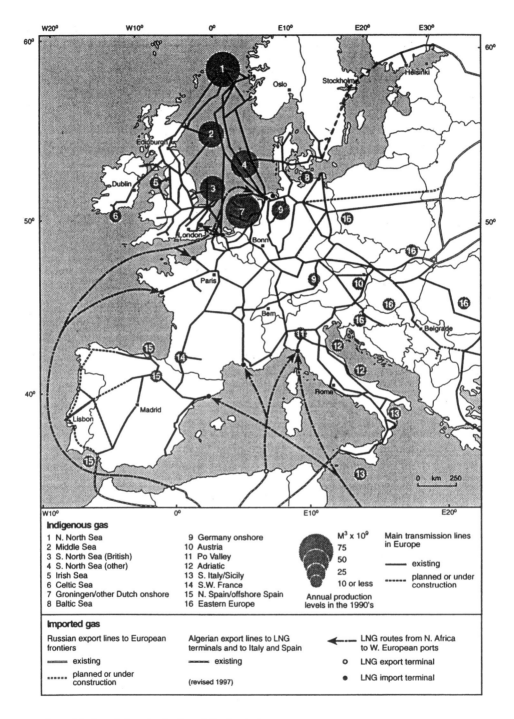

Figure 4.4 Europe's Emerging Continental Gas Supply System

through Russia or Turkey from Azerbaijan, Turkmenistan and Kazakhstan (Roberts, 1998; *Oil and Gas Journal*, 2002). The location of the latter countries together with Uzbekistan is, however, difficult, compared with that of Russia, as it involves higher transport costs, partly because of the distance and partly because of the imposition of transit fees by the intervening countries (Estrada, 1995; Wybrew-Bond and Stern, 2002). There are, indeed, already indications that the former Soviet republics in Central Asia could more appropriately seek links with south-east Asia and even east Asia. In south-east Asia, however, their gas export potential will have to compete with supplies from the Middle East; while in east Asia their efforts will run into competition from Russia's large gas potential in East Siberia and Sakhalin. The gas export potential of these Soviet-created and now somewhat artificial Central Asian republics, with a combination of political and geographical disadvantages for foreign trade, currently seems unlikely to be achieved on a large scale before the second quarter of the 21st century (*Oil and Gas Journal*, 2002; Holmes, 2003).

Nevertheless, in the long-term the prospective role of the hydrocarbons-rich countries of the FSU in Eurasia's gas markets will be achieved on the basis of their large share of the world's conventional and non-conventional gas resources (see Tables 4.1 and 4.2): together with the potential post-2050 exploitation of some small part of the quarter of the world's ultimate resources of gas from hydrates which are estimated to lie within their boundaries and offshore waters (see Table 4.2). Indeed, the gas wealth of Russia and of the other gas-rich former Soviet republics provides what can perhaps best be described as the single most significant element in the world's prospective involvement with carbon fuels from 2050 onwards.

The Middle East

This region's dominance in conventional oil reserves is not matched by its reserves of conventional gas. It is, nevertheless, still an immensely gas-rich area with proven reserves of 47Gtoe, second only to the FSU with 51Gtoe (see Table 4.1). Geographically, the Middle East's reserves are much more heavily concentrated as they lie in an area only about 8% of the area of the FSU. This contrast is not insignificant for development potential, given the relatively high costs of gas transportation. However, in terms of the cumulative production of gas, the FSU has produced more than eight times that of the Middle East, while in terms of year 2000 gas production the FSU's was more than three times that of the Middle East. Likewise, the FSU's estimated additional reserves (96–110Gtoe) are significantly more substantial than those of the Middle East, viz. 29–50Gtoe as also shown in Table 4.1. The contrast continues in the estimates in Table 4.2 of non-conventional reserves excluding gas hydrates (159Gtoe in the FSU against 99Gtoe in the Middle East). For gas hydrates the potential in the FSU could be up to 20 times that of the Middle East. Compared with its unique position in respect of oil reserves, the Middle East's natural gas future is significantly restrained not only by this greater FSU potential, but also by their contrasting political relationships with potential importing countries of Europe and Asia.

These contrasts in the political, economic and social characteristics between the Middle East and the FSU serve to make the two regions much more competitive in respect of their export potential. While gas use in the Middle East remains at only a third of that in the FSU, this contrast will become somewhat less pronounced in the coming decades. It is, however, unlikely to fall below a ratio of 5:2. Thus the Middle East could become a potential rival to the FSU, in general, and to Russia, in particular, in respect of gas exports to accessible regions which have inadequate resources to meet their future gas needs. Europe has been designated above as one such region, but it is already importing over 128 Bcm (0.12Gtoe) of Russian gas annually. Moreover, the political changes in the FSU in the 1990s have eliminated previous political restraints on Europe's willingness to expand its Russian gas imports, with the result that major infrastructure developments for much enhanced gas deliveries are under way (Bakhtiari, 2003). The same is not yet true of the Middle East. Its gas exports to Europe remain close to zero (viz. 4Bcm of liquefied natural gas in 2002) and no pipeline connectors are in prospect. Large-scale gas supplies from the Middle East to Europe remain a long-term prospect only (Odell, 1995 and 2002; Gas Matters, 2002).

Future use of the Middle East's gas reserves seems more likely to be orientated to south-east Asian markets, notably the Indian sub-continent, where proven and even additional gas reserves are modest, relative to potential demand. Economic and political considerations seem likely to inhibit links between the Middle East and Asia for the short and medium term, but in the long term the complementary relationship of the two areas – with large gas reserves and resources, on the one hand, and potentially large gas markets, on the other – will exercise sufficient force to overcome the constraints, although still in competition with possible large-scale flows of gas from the former Soviet republics of Central Asia (Roberts, 1998). Meanwhile, Middle

East exports of LNG will build-up quickly from their current level of 35Bcm per year as major liquefaction projects in Qatar, Saudi Arabia, Oman and Iran are completed (Flower and King, 2002). By 2020 these could supply upwards of 100 Bcm, principally serving North American and East Asian markets to which deliveries by pipelines will be constrained by geography. The coastal or near coastal location of most of the Middle East's abundant gas reserves and potential gives the region a 'natural' advantage of significant proportions for LNG production facilities, compared with the higher transport costs of gas from the generally remote inland location of Russian and Central Asian reserves. Mid- to late-21st century LNG exports from the Middle East of up to 250 Bcm are not beyond the limit of expectations (Jensen, 2003).

Africa
The Mediterranean littoral states, with almost 75% of Africa's proven gas reserves and an estimated 50% of the continent's unexploited and undiscovered conventional gas reserves (USGS, 2000), have long been considered as part of the Eurasian gas region (Estrada, 1995; Odell, 1995). Algeria already competes with Russia for exports to Europe and expansion of the volume traded is certain as new trans-Mediterranean pipelines are completed (Mabro and Wybrew-Bond, 2001). Libya and Egypt will join the system within the next 10 years. Whilst long-term resource availabilities will certainly sustain production expansion of significant dimensions, there will be constraints on the volumes of exports to Europe within 20 years, as demand growth there falls away with the satiation of markets and in the context of competition from other suppliers (see above pp.76–7). Thus there seem likely to be limits to the ultimate scale of the production developments in the countries of North Africa. Alternative LNG exports to the United States will have to provide the required longer-term outlets, but in competition with supplies from Atlantic basin sources, such as Trinidad, Venezuela and Nigeria (Jensen, 2003; Quinn, 2002).

One country, Nigeria, with about 3600Bcm (=3.2Gtoe) of gas reserves dominates the currently proven gas reserves of the rest of Africa (with only 600 Bcm). Its large reserves have been known for 25 years, but economic and political conditions in the country as well as its location which allows gas to be exported only as LNG, has kept the exploitation of its large gas reserves on a low plateau for almost a decade. A large, almost 20-year-old LNG export project was finally completed in 1999. Other such LNG export projects are now under development and will more than double the volumes which can be exported to European and North American markets. Meanwhile, Nigeria's considerable gas reserves, plus smaller volumes of gas in other countries of West Africa, could form the base for a gas transmission system in that part of the continent, but such a capital-intensive and politically difficult system in an area of low energy demand remains a long-term prospect only. In the longer-term, such regional pipeline developments could provide the basis from which a long mooted trans-Saharan line could eventually be built; though cooperation with Algeria in this development may well prove difficult to achieve, given that Nigerian gas would then become a competitor in Europe for supplies from North Africa. An African gas-grid remains no more than a very long-term possibility.

In spite of limited exploration efforts elsewhere in Africa, significant gas reserves (in relation to the small energy markets) have been discovered, but are not yet being much exploited for exactly the same reasons as noted above for Nigeria. South Africa's high and growing demand for energy provides a small market for gas from Mozambique, but the low cost of exploiting South Africa's rich coal reserves does constitute a limitation on gas markets' developments, especially for power generation. Similarly, offshore gas from Namibia has to date failed to generate interest enough in South Africa to make its exploitation profitable. Again, prospects for exploitation remain only for the long term, and, even then, will be on a modest scale by world standards. Angola's growing gas reserves – mostly as gas associated with its large-scale offshore oil developments only seems to be exploitable as LNG for export to the United States – in competition with supplies from other more favourably located LNG projects (*Oil and Gas Journal*, 2003a).

Geologically, the Great Rift Valley of East Africa remains an enigma in respect of its gas potential. Limited exploration to date has proved the existence of modest reserves, but there remains a possibility of major discoveries from an intensified effort. Given its location, however, in terms of both regional markets and export potential, the region will remain low in the international order of priorities of most companies.

In brief, except for the Mediterranean-orientated countries and Nigeria, the continent seems highly likely to remain unimportant in global gas industry development terms for at least the first quarter of the 21st century. Thereafter, Africa could become the last of the world's continents (except for Antarctica) to secure extensive access to natural gas and, even more so, to make possible its intensive use.

Asia Pacific

This dominant world region, in terms of population and of prospective economic growth, is not only relatively poor in currently declared proven gas reserves – with only 7% of the global total – but is also estimated to have relatively limited potential for additional reserves of conventional gas (Table 4.1). The region's clear-cut bottom ranking for the contribution of gas to total energy use (Table 4.3) seems, moreover, to reflect a more fundamental problem than the historical lack of interest in exploiting the region's gas resources. Even in the 1990s, at a time of a rapid increase in the region's energy use, gas' share in the energy market rose only from 7 to 10%.

This generally modest role of natural gas in the region did not, however, entirely exclude some important developments in the recent past in the exploitation of the region's known gas reserves (which provide a reserves-to-production ratio of over 41 years). Gas production increased by almost 100% in the 1990s, based largely and uniquely on the innovative LNG production and transport technologies that have been implemented as a result of the archipelagic character of the region. Indeed, by 2000, over two-thirds of international trade in LNG was to Asian Pacific markets – with no less than 80% of this originating from supplying countries within the region (Flower and King, 2002).

The high costs involved in the creation of a comprehensive and integrated gas pipeline system – in a region in which markets are fragmented by physical geographical characteristics – remain a barrier to pipelined gas being made available to users, and hence to the more intensive exploitation of the actual and potential resources of the regions. Plans for an international pipeline system within the region have been proposed and discussed by regional development organisations, but implementation has as yet been only modest, for example, from Indonesia to Malaysia and Myanmar to Thailand.

Thus, as with oil (see Chapter 3), major and continuing expansion of gas use for the period to 2020 is more likely to depend on imports from outside the region, viz. from the Middle East into the Indian sub-continent; from Russia's Far Eastern gas potential into China, Japan and Korea; from the Central Asian republics (Turkmenistan and Kazakhstan) to India and eventually by a long distance line to China (Paik, 1995; Gas Matters, 2001; IEA, 2002b; Wybrew-Bond and Stern, 2002). For the smaller economies of south-east Asia which are relatively better endowed with gas and already more dependent upon it, the more intensive developments of indigenous resources and the near-future construction of more pipeline connections between neighbouring countries – such as Indonesia to Singapore, Papua New Guinea to Australia and Bangladesh to India – will gradually be achieved and keep that part of the region's gas economy growing faster than its energy economy in general. Meanwhile, both China and India – the region's largest countries and economies by far (except for Japan) – have enhanced and accelerated their gas exploration efforts; with potentially significant results for the development of much increased indigenous gas production and consumption by the second decade of the 21st century (Chandra, 2003; Chow, 2003; Gas Matters, 2003b).

The region's major LNG producers, viz. Australia, Brunei, Indonesia and Malaysia, will also continue to build-up their LNG exports to Japan, South Korea, Taiwan and, even more significantly in the future, China. This, in part, will reflect these energy importing countries' attempts to limit their dependence on Middle East oil. Somewhat ironically, moreover, Middle East oil will also be substituted by LNG from the Middle East – notably Qatar, Oman and the United Arab Emirates (Quinn, 2002; Wybrew-Bond and Stern, 2002)

There remains one extraordinarily important unresolved geopolitical dispute in the Asian Pacific region which relates especially to natural gas, viz. the issue over the sovereignty of the South China Sea, in general and, in particular, to the disputed ownership of the Spratly Islands, a small group of tiny islands in a huge, relatively shallow water area which is considered to have high hydrocarbon potential (Figure 4.5). The claim by China to the exclusive mineral rights over most of the South China Sea affects five south-east Asian countries and has effectively stymied the effective petroliferous exploration of the area; from which it is expected significant reserves of gas could be produced and transmitted in conventional relatively short distance pipelines to the region's centres of demand (Paik and Kim, 1995). The medium to longer-term prospects for the development of these natural gas resources for the countries concerned depend significantly on a solution to the political problem.

Significant New Developments for Gas' Expansion

First, natural gas-fuelled combined cycle power generation has recently come to provide the basis for a much enhanced requirement for gas supplies in the United States and Europe. This reflects the technology's high conversion efficiency so enabling higher cost gas to compete effectively with the use of lower cost coal. The development also serves energy policies which aim to reduce CO_2 emissions (Haites and Rose, 1996; Freund, 2002). This technical development, with its attractive economic and environmental advantages, will be largely responsible for the more rapid exploitation of known gas reserves. It will enhance the rate of exploration for new reserves – in an increasing number of countries in all parts of the world – on a continuing basis in the first half of the 21st century.

Second, another potentially equally significant technical development which will enable gas to compete with oil products in the latters' most important remaining markets is also now getting under

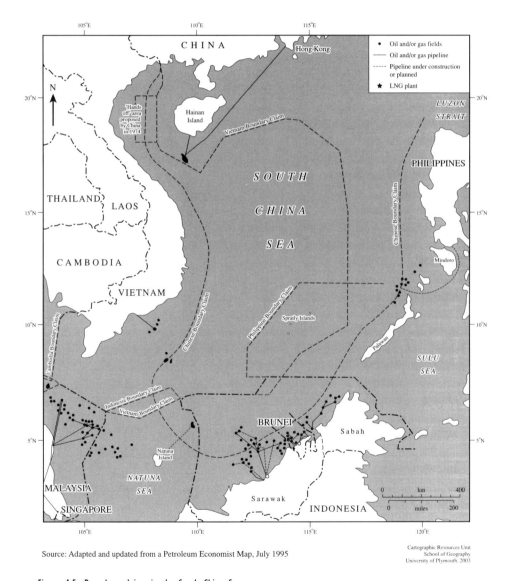

Source: Adapted and updated from a Petroleum Economist Map, July 1995

Cartographic Resources Unit
School of Geography
University of Plymouth, 2003

Figure 4.5 Boundary claims in the South China Sea

way, viz. the conversion of natural gas into liquid transportation fuels (GTL) (Shook, 1997; Thackeray, 2002). There is a huge market potential for gas from this development as transportation fuels now account for over 50% of oil use. The volume of oil thus used is roughly the equivalent of two-thirds of the present total world market for natural gas. GTL technology also fundamentally alters the economics of geographically remote gas exploitation, hitherto inhibited by the high costs of getting such gas to market. Oil to gas conversion plants in resource-orientated locations – including offshore locations using floating conversion facilities – will overcome this problem. Additionally, the liquid products of GTL conversion yield cleaner-burning fuels, with lower emissions of both particulates and greenhouse gases. The medium to longer-term impact of GTL on gas use – and hence on the future geographical spread and intensity of the exploitation of the world's gas reserves, eventually including non-conventional gas occurrences – will thus be a major factor in ensuring the rate of build-up of gas supplies as indicated earlier in this chapter. It thus confirms the practicality of the long-term process of substituting gas for oil, as discussed in Chapter 3.

Third, there is also a potentially equally significant – albeit an even longer-term – role for gas in the transport sector. There are two sequential elements to this development. In the shorter-term there will be a major expansion of compressed natural gas (CNG) as an alternative fuel for road transport vehicles – as a direct substitute for oil products. This has already made a not-insignificant contribution to fuelling specific types of vehicles, notably buses and delivery vans operating within a limited range of their depots/garages, so avoiding the disadvantage of CNG arising from the need for the more frequent re-fuelling of the vehicles. Technical developments to overcome this problem – and the eventual mass production of gas-using engines – will, in time, expand the percentage of vehicles which can run on natural gas.

Longer-term, enhanced concern for the atmospheric pollution and the high emissions of greenhouse gases by motor vehicles will likely stimulate requirements for motor transportation based on the use of hydrogen. In the context of the need to produce that hydrogen at as low a cost as possible, the only foreseeable means would be its production in large-scale static plants using natural gas as the input fuel (and with provision for the CO_2 by-product to be collected for subsequent sequestration, rather than its release into the atmosphere) (Griffiths, 2001; Hoffman, 2001; Rifkin, 2002). This would add significantly to the global demand for natural gas at a time when non-conventional gas production will, as shown in Figures 4.1 and 4.2, be emerging as the most important component in the commodity's long-run supply curve. The rise in gas output to an annual rate of more than 11Gtoe by 2090 (over 50% higher than the peak rate forecast for oil production) is thus predicated in part on a gas conversion-to-hydrogen requirement whereby a more environmentally friendly global energy economy can be created.

References

Adelman, M.A. and Lynch, M.C. (2003), *Natural Gas Supply to 2100*, Hoersholm, International Gas Union.

Baktiari, A.M. (2003), "Russia's gas production and exports." *Oil and Gas Journal*, Vol.101.10.

Bartsch, U. and Müller, B. (2000), *Fossil Fuels in a Changing Climate*, Oxford, Oxford University Press.

Bradley, R.L. (2003), *Climate Alarmism Reconsidered,* London, Institute of Economic Affairs.

Canadian Gas Potential Committee (2001), *Natural Gas Potential in Canada*, Calgary.

Carroll, J. (2003), *Natural Gas Hydrates*, Burlington, Mass, Gulf Professional Publishing.

Chandra, A. (2003), "Sizeable finds from India's licencing efforts," *Oil and Gas Journal*, Vol.101.23, pp.41–4.

Cherkashov, G.A. and Solawiev, V.A. (2002), "Economic use of hydrates: dream or reality." *Proceedings of the 17th World Petroleum Congress*, Vol.1, pp.253–62.

Chow, L. (Ed.) (2003), "Themes in current Asian energy." *Energy Policy*, Vol.31.11.

Cornot-Gandolphe, S (1995), "Changes in world natural gas reserves and resources," *Energy Exploration and Exploitation*, Vol.13.1, pp.3–18.

Delahaye, C. and Grenon, M. (Eds.) (1983), *Conventional and Unconventional World Natural Gas Resources*, Laxenburg, IIASA.

Energy Information Agency (1994a), *Energy Use and Carbon Emissions: some International Comparisons*, Washington, DC, US Government Printing Office.

Estrada, J. et al (1995), *The Development of European Gas markets*, Chichester, J. Wiley & Sons Ltd.

Flowers, A. and King, R. (2002), *LNG today: the Promise and the Pitfalls*, Oxford, Energy Publishing Network.

Freund, P. (2002), "Technology for avoiding CO_2 emissions," *Proceedings of the 17th World Petroleum Congress*, Vol.5, pp.11–21.

Gas Matters (2001), "World gas majors involvement in cross-China pipeline project," London, August.

Gas Matters (2002), "Iran rekindles European gas export ambitions," London, February, pp.11–15.

Gas Matters (2003), "Coal-bed methane creeps up the alternatives chart," London, August, pp.15–21.

Gas Matters (2003b), "Can India show enough market potential to lure new gas supplies," London, August, pp.3–10.

Gregory, K. and Rogner, H-H (1998), *Energy Resources and Conversion Technologies for the 21st Century*, Vienna, International Atomic Energy Authority.

Griffiths, J. (2001), "Fuel cells – the way ahead," *World Petroleum Congress Report*, London, ISC Ltd.

Haites, E.F. and Rose, A. (1996), "Energy and greenhouse gas mitigation," *Energy Policy Special Issue*, Vol.24.10/11, pp.857–1016.

Hoffman, P. (2001), *Tomorrow's Energy: Hydrogen, Fuel Cells and the Prospects for a Cleaner Planet*, Cambridge, Mass, Cambridge University Press.

Holmes, C. (2003), "Uncertainties for FSU projects threaten export potential," *Oil and Gas Journal*, Vol.101.22.

International Energy Agency (2002a), *Flexibility in Natural Gas Supply and Demand*, Paris, OECD.

International Energy Agency (2002b), *Developing China's Natural Gas Market*, Paris, OECD.

International Energy Agency (2003b), *World Energy Investment Outlook*, Paris, OECD/IEA.

International Gas Union (1997), "World gas prospects: strategies and economics," *20th World Gas Conference*, Copenhagen.

International Gas Union (2003), "Catalysing an eco-responsible future," *21st World Gas Conference*, Tokyo.

I.I.A.S.A. (1995), *Global Energy Perspectives to 2050 and Beyond*, London, World Energy Council.

Jean-Baptiste, P. and Ducroux, R. (2003), "Energy policy and climate change," *Energy Policy*, Vol31/2, pp.155–66.

Jensen, J.T. (2003), "The LNG revolution," *The Energy Journal*, Vol.24.2, pp.1–46.

Kurtz, D. (1997), *Natural Gas in Latin America: Development and Privatisation*, London, Financial Times Business.

Lerche, I. (2000), "Gas hydrates," *Energy Exploration and Exploitation*, Vol.18.2 and 3.

Lerche, I. (Ed.). (2002), "Gas in the 21st Century," *Energy Exploration and Exploitation,* Vol.20, No.4.

Lowrie, A. and Max, M.D. (1999), "The extraordinary promise and challenge of gas hydrates," *World Oil*, September, pp.49–55.

Mabro, R. and Wybrew-Bond, I. (Eds.) (1999), *Gas to Europe: the Strategies of the Major Suppliers*, Oxford, Oxford University Press.

Marchetti, C. (1978), *Energy Systems: the Broader Context*, Laxenburg, IIASA.

Moritis, G. (2003), "CO_2 sequestration adds new dimension to oil and gas production," *Oil and Gas Journal*, Vol.101.9, pp.39–44.

Norwegian Ministry of Petroleum and Energy (2002), *Report on the Norwegian Continental Shelf*, Oslo.

Odell, P.R. (1969), *Natural Gas in Western Europe*, Haarlem, de Erven F. Bohn.

Odell, P.R. (1981), "Prospects for and problems of the development of oil and gas," *Natural Resources Forum*, Vol.5.4, pp.317–26.

Odell, P.R. (1984), *The Oil and Gas Resources of the Third World*, New York, United Nations Development Division.

Odell, P.R. (1988a), "The western European gas market – the current position and alternative prospects," *Energy Policy*, Vol.16.5, pp.480–93.

Odell, P.R. (1992), "Global and regional energy supplies – fictions and fallacies," *Energy Policy*, Vol.20.4, pp.284–96.

Odell, P.R. (1995), "The cost of longer-run gas supply to Europe," Stockholm, SNC Reprint paper no.25.

Odell, P.R. (1998), "Oil and gas reserves: retrospect and prospect, *Energy Exploration and Exploitation*, Vol.16.2, pp.117–124.

Odell, P.R. (2002), *Oil and Gas: Crises and Controversies, 1961–2000*, Vol.2, Brentwood, Multi-Science Publishing.

Oil and Gas Journal (2002), "Focus on Turkmenistan", Vol.100, Nos.41–44.

Oil and Gas Journal (2003a), "Gas in the US," Vol.101.10, p.19 and Vol.101.11, p.23.

Paik, K-W. (1995), *Gas and Oil in North East Asia: Policies and Prospects*, London, R.I.I.A.

Paik, K-W. (2002), "Natural gas expansion in China," *Geopolitics of Energy*, Vol.24, No.5.

Paik, K-W. and Kim, D-K. (1995), "The Spratly islands' dispute with China's naval advance", *Geopolitics of Energy*, Vol.17, No.10, pp.5–10.

Quinn, A.C. (2000), "Long-term LNG contracts to opportunity markets," *Proceedings of the 17th World Petroleum Congress*, Vol.4, pp.185–192.

Rifkin, J. (2002), *The Hydrogen Economy*, London, Penguin Books.

Roberts, J. (1998), "Gas from the Caspian", *Geopolitics of Energy*, Vol.20, No.5, pp.1–3.

Shell IPC (2001), *Energy Needs: Choices and Possibilities: Scenarios to 2050*, London.

Stinemetz. D. (2003), "Russian oil and gas sector rebound in full swing," *Oil and Gas Journal*, Vol.101.22, pp.20–30.

Shook, B. (1997), "Gas to liquids emerges from the fringe," *World Gas Intelligence*, December 19, pp.7–10.

Thackeray, F. (1998), "The future belongs to gas", *Petroleum Review*, March, pp27–29.

Thackeray, F. (2002), "The promise of gas-to-liquids technology," *17th World Petroleum Congress Report*, London, ISC Ltd. pp.176–83.

Torp, T.A. (2001), "Carbon sequestration: a case study." *17th World Petroleum Congress Report*, London, ICS Ltd., pp.156–9.

United States Geological Survey (1997), *World Energy Resources*, Washington DC.

United States Geological Survey (2000), *World Petroleum Assessment*, Reston, Government Printing Office.

United States Geological Survey (2001), *Natural Gas Hydrates: Vast Resources, Uncertain Future*, Reston, Government Printing Office.

Van de Vate, J. (1997), "Comparison of energy sources in full chain emissions of GHG," *Energy Policy*, Vol.25.1, pp.1–6.

Vyakhirev, R.I. (1997), "Natural gas in Russia: potential for the 21st Century," *Proceedings of the 15th World Petroleum Congress*, Forum 10.

World Petroleum Congress (2002a), "New hydrocarbon provinces of the 21st century," *Proceedings*, Vol.2, Forum 2.

World Petroleum Congress (2002c), "Natural gas: clean energy for half-a-century," *Proceedings*, Vol.4. Forum 14.

Wybrew-Bond, I. and Stern, J. (2002), *Natural Gas in Asia*, London, R.I.I.A.

Chapter 5: Trends in Production Costs and Prices

Introduction

In dealing with each of the carbon fuels, both specific and implicit references have been made to trends in costs and prices, but these observations need to be brought together into an integrated overview. Questions of costs and, even more so, of returns on investments constitute a general set of parameters relevant to most resource evaluations, in the context of the existence of a common measure of economic accounting in the global system. Few significant differences can emerge between the prices of fuels except as temporary phenomena, usually arising either from national policies involving protection for indigenous production, or from major disruptions to international trade caused by wars (Odell, 1971; Heinberg, 2003) and revolutions (Clawson, 1964). Except in such extreme circumstances, then were such divergences to emerge, they would inevitably lead to compensating changes in supply schedules so that the equilibrium could be re-established. But, the restoration of equilibrium is inevitably a time-consuming process so that there may occasionally be a need for subsidies or other protective measures in specific markets to encourage the production of more expensive resources; as in the aftermath of the oil price shocks of 1973/4 and 1979/80 (Odell, 1986). In general, of course, contrasting transportation costs in getting production to markets produce contrasting netbacks for different fuels in the same region and between regions for the same fuel, but in a macro-study of the overall global situation over a long period of time, such transport orientated differences are unlikely to mount to much more than noise in the structure of the global energy system.

The starting point of this long-term study of the future availability of coal, oil and natural gas is the declaration of proven reserves for each fuel. Such declarations are, by definition, related to reserves that are economic to produce at current levels of costs and prices. As the total quantity of the reserves declared is much more than adequate to serve the world's slowly expanding markets, in the context of price levels which through most of the 20th century were higher than the long-run supply price (Odell, *ibid*.), then the outcome of competition between energy sources and between suppliers is to keep prices from escalating in real terms. Of course, traumatic events, like those in the past – such as US oil import quotas in the 1960s and 1970s and the supply limitations imposed by OPEC and others in the 1970s and 1980s – have seriously upset the equilibrium from time to time (Fisher, 1974; Schurr and Netschert, 1977; Adelman, 1993). Similar events will undoubtedly occur in the future, but they cannot be forecast and therefore cannot be taken into account in an attempt to predict overall long-term cost and price trends.

Supply Costs for Carbon Fuels

In spite of high consumer energy prices having been the norm, rather than the exception, over the past 30 years, supply costs of coal, oil and gas have generally remained low (Adelman, *ibid*.; Hartshorn, 1993) as increasing demands for the fuels have been more than matched by continuing growth in supply and supply potential. The world's carbon energy resources have thus gradually been shown to be plentiful, while technological developments have reduced discovery and exploitation costs of the resources (Rees, 1985; Gordon, 1987; Hartshorn, 1993; Alazar, 1996; McCabe, 1998).

Notwithstanding some claims that the era of low-cost carbon energy has already, or will soon, become out-of-date (Edwards, 1997; Campbell, 1997; Campbell and Laherrère, 1998; Laherrère, 2003), the evidence suggests otherwise. Through the increasingly effective application of advancing technology and developing managerial skills, investments have become more productive in creating new and expanding capacity in the industry (Econ Centre for Economic Analysis, 1993). This process has been evident, for example, over the relatively short history of the development of offshore conventional oil and gas resources (Smith and Robinson, 1997; Norwegian Ministry of Petroleum and Energy, 2002); and, more recently in the exploitation of non-conventional heavy oils and tar sands in Alberta. Here costs per unit of production have fallen by upwards of 50%, to ±$10 per barrel within a decade (Lang et al, 2000; OGJ, 2003b). More generally, finding and development costs of oil in non-OPEC countries have fallen from $22 per barrel in 1981 to $6 per barrel in 2001 (Baird, 2003).

Thus, recent history demonstrates that the long-run supply price curve for carbon energy has not yet turned up to any significant degree. That this is not likely to happen for the first 30 years of the 21st century is, moreover, confirmed in a major new study on global energy sector investment costs published by the International Energy Agency (IEA, 2003b). This work also incorporates inputs from other relevant parties including OPEC, the World Bank and energy companies. It forecasts the investments which will be needed to secure a 65% increase in world energy supply by 2030: an expansion which is somewhat higher than that forecast in this study (see Figure 1.5).

First: for coal, capital investments of $400 billion ($10^9$) are estimated as being required for the exploitation of about 90Gtoe (= 135 billion tons of coal) over the 30 years to 2030. This makes for a very modest average of $2.96 of investment in the production and transportation facilities for each ton of coal going into the market. This is the equivalent of only $0.61 per barrel of oil.

Second: for oil, a very much larger total capital investment of $3100 billion is calculated to be required to build the infrastructure needed to discover, produce and transport and refine 140Gtoe of crude oil (over 1000 billion barrels). Even so, the investment per ton is still only $22.15; equal to just over $3 per barrel.

Third: for natural gas, the estimate of the required investment is also $3100 billion, but in this case for the exploitation of 95Gtoe (= 105 x $10^{12}m^3$). Of this total amount $1700 billion is required for the exploration and exploitation infrastructure and $1400 for the transmission and distribution of the gas inclusive of LNG liquefaction, tanker shipments and regasification terminals. This investment works out at $32.65 per ton or $4.45 per barrel of oil equivalent (= 2.9 US cents per cubic metre and to just under $1 per million Btus).

These investment costs estimates for carbon energy infrastructure developments show that the capital components in future supply costs (in year 2000 dollar values) are far below current price levels. Even if one generously assumes that the addition of the industries' operating costs and of the profit margin which the suppliers will need to secure, together add 200% to the investment costs as defined above, viz. to $9 for the full cost of a ton of coal delivered to market; to just under $10 for a barrel of refined oil; and to under 9 US cents per cubic metre of gas (or approximately $2.70 per million Btus), then the continuation of present price levels for all carbon energy sources would generate super-normal profits for the companies involved and/or significant taxation opportunities for the governments of the producing/exporting countries.* The arguments to support a very long-term future for low-cost carbon energy prices is set out below in the context of these preceding considerations.

Prices to 2040
The potential supply schedule we have forecast for oil (See Figure 3.4) – for long the price leader in the carbon fuels' market – suggests little or no pressure of demand on supply for at least the next 20 years. This is confirmed by the results of the IEA study as discussed above. There is thus no reason why prices, currently will above the long-run supply price of coal, oil and natural gas, should rise in real terms and no reason why any significant volumes of reserves of oil (or gas or coal), which involve much higher costs, will need to be produced. The interquartile price range over the final fifteen years of the 20th century for internationally traded crude oil (calculated monthly by the US Department of Energy) was $18.20 to $22.35 per barrel – valued in year 2000 dollars. This still represents a generally acceptable price range for oil – also in year 2000 $ values – under the supply/demand conditions that have been predicated in this study for conventional oil: and for the smaller volumes of unconventional oil and gas which will be demanded to 2020 (see Figure 1.5 and also Chapter 3). The likely maintenance of this level of prices in the market place for the first two decades of the 21st century is also related to the fact that it lies within the range of prices which OPEC seeks to achieve, viz. $20 to $26 per barrel in year 2000 dollar values and is in line with the "preferred" price for oil of many other governments. Likewise, the oil companies have a vested interest in prices at much the same level (Odell, 2001).

In this context of an "ordered" market as described above, and which seems highly likely to persist in the same way that it did for most of the 20th century (Adelman, 1972; Odell, 2001), the price of

---

* This set of simplistic calculations based on the averaging of the global financing requirements for three decades does not, of course, recognize either the spatial or the temporal variations in the capital costs of producing and delivering the volumes of coal, oil and gas as and when required and in the context of a market economy in which prices will reflect the costs of the marginal supplier. Nevertheless, the aggregated and the averaged costs derived from the I.E.A. study provide important evidence in support of our hypothesis for a prospective long-continuing low-cost global carbon energy economy.

oil is not in much danger of falling back closer to its long-run supply price of $10 to $15 per barrel (Odell, 1986; Adelman 1993). Higher prices than $26 per barrel can, on the other hand, only be expected from time to time over this period in the context of supply disruptions arising from radical political changes or military conflict. The latter will principally be in the Middle East from which supply area the largest volumes of exports to international (as opposed to regional) markets will continue to flow (Odell, 1997a).

After 2020, however, upward price pressures on the oil market are possible as the efforts to maintain growth in the final decade of increasing global production of conventional oil (see Figure 3.4) will necessarily lead to rising investment costs. The need for an increase in prices to offset the 10–20% increase in the exploration/development costs for conventional oil will then be more than confirmed by the higher unit investment costs involved in the substantial exploitation of non-conventional oil on a large and continuing scale in locations other than Western Canada and Venezuela. Such cost increases will be passed through into the general level of prices. Thus, the equilibrating $18–22 per barrel of oil in the meantime, seems likely to be converted by 2020 to $21–26 per barrel (in year 2000 dollar values). At this higher price level, the requirement for the highest cost oil producer to sell into the market and earn sufficient profits will be satisfied.

Neither gas nor coal suppliers are likely to need the price increase as plentiful supplies at pre-existing costs will still be available but, with oil prices still essentially setting the general price level in the global energy market, the coal and gas producers will enjoy an enhanced economic rent – while the producing/exporting countries secure increased tax payments. In the case of coal, increases in the price of oil and gas will also enable the industry to absorb the relatively higher carbon taxes to which much coal production by then seems certain to have become subject. Equilibrium will therefore be maintained between the three carbon energy sources in a situation in which alternative energies will, in general, still require a production subsidy and thus be unable to bring downward pressure to bear on energy prices by that time.

Mid-century Prices

A likely near mid-century price crunch (when upward pressures generated by relative scarcity exceed the downward pressures engendered by technology), will emerge as conventional gas output approaches its period of maximum production in the 2040s (see

Figure 4.1). By then gas is indicated to have overtaken oil as the world's single most important source of energy (see Figure 3.6). It will have achieved a much broader market structure, having replaced much coal in the power generation market and large volumes of oil products in the transportation markets. In essence, natural gas will by then have become the energy market leader. But, in order to sustain the growth of the industry by means of the profitable exploitation of non-conventional gas, a further price increase (in real terms) will be required to generate the higher level of investments which will be required for the exploitation of non-conventional gas. This could take the oil equivalent price to a level of $25–30 per barrel (in year 2000 dollar values).

By that time, however, yet higher taxes on the use of coal for environmental reasons will be tending to price the fuel out of many of its hitherto main markets (see Table 2.5). Technological improvements in coal production notwithstanding, coal will no longer have much ability generally to influence energy prices, except for power generation. The influence of coal on energy prices by then thus seems likely to be restricted to exercising downward pricing pressures in a limited number of coal-rich countries.

Oil will also have become largely a price taker – rather than a price maker – so that the industry will be able to benefit from the higher revenue flows generated from the upward movement of gas prices. It will thus secure adequate returns on the large investments required for non-conventional oil production between 2040 and 2060 – the period of the most rapid rate of growth for this source of oil.

Likewise, the by-now expanding delivery of renewable energies will benefit from the upward price movements in the market. The higher prices will enable governments to reduce, or even eliminate, the subsidies hitherto required for sustaining the growth of the industry.

From 2040 to 2060 the supply of energy, in general, and of carbon energy, in particular, will continue to become increasingly orientated towards natural gas as the industry continues to expand its markets. It will thus be motivated and required rapidly to enhance the output of non-conventional gas as conventional gas supplies reach their peak in 2050. This mid-21st century period will need to be the age in which gas supply and users' technologies are perfected, so that investment and other costs do not rise significantly in real terms in the context of an average annual growth rate in supply of about 2%. Gas prices will also be under pressure to remain stable in real terms because of increased competition from subsidised and hence economically attractive renewable energy supplies. The intensification of the supply of renewables will by then begin to make large economies of scale possible for the first time in the history of the non-carbon fuels energy industry.

Prices Post-2060
For the rest of the century from 2060 we have predicated that increases in the supply of gas at an annual growth rate of 2% will no longer be needed. This is because the rate of increase in global energy demand overall will, by then, be falling away in the context of a slower rate of increase in the world's population. Fortuitously, the gas industry at that time will be moving towards its peak production in 2090, comprising approximately one-third conventional and two-thirds non-conventional gas. Thus, the cost increases required to maintain gas production at the high global level of 11.2Gtoe then achieved (compared with a peak production of oil of 6.6Gtoe 30 years previously) may well become a major issue as the technological improvements on the supply side reach their limit. Whether or not such cost increases will become significant in the early years post peak production in 2090 cannot be assessed, but even if they do there is no certainty that it would be possible to pass them on in higher gas prices. This would be a consequence of the improving economics and technology of energy supplied by renewable sources so that the latter could by then be setting the price levels. Nor is it impossible that the large resources of low-cost coal remaining in the late 21st century, as a consequence of the restrained use of only some 215Gtoe of coal between 2000 and 2060, may finally be exploited at a significantly lower cost and thus become an appropriate replacement fuel. This could arise either from the failure of global warming to persist (so that CO_2 emissions are no longer considered a problem), or from the emergence of technologies (such as clean coal combustion and CO_2 sequestration) which would enable coal to be used in an environmentally friendly way.

An Overview of Developments
There thus appear to be only two periods in the 21st century when fuel costs and prices are likely to rise in real terms; viz. first, towards the end of the second decade of the century, because of the then approaching need for conventional oil production to be supplemented post-2020 by inherently higher cost non-conventional oil production, in a situation in which oil still remains the world's most important

energy source; and second, around 2040, when non-conventional gas production has to be commercialised on a large scale, at a time when renewable energies have not yet become capable of taking over the responsibilities for substituting more expensive gas.

After 2040, the changing combinations of flows of different sorts of energy, to serve the soon-to-emerge slow but steady fall in the global energy demand growth rate, seem likely to emerge quite smoothly in the context of the higher real prices then achieved. Towards the end of the century, in the context of continuing technological progress, a secular fall in the real costs of supplying renewable energy seems more likely than rising costs. If so, then the phenomenon will serve to bring downward pressure to bear on the prices of carbon fuels. Such a development could then conceivably accelerate their late-21st century declining contribution to world energy supply, as they become progressively less able to compete effectively in the global energy market-place.

References

Adelman, M.A. (1972), *The World Petroleum Market*, Baltimore, The John Hopkins University Press.

Adelman, M.A. (1993), *The Economics of Petroleum Supply*, Cambridge, Mass, MIT Press.

Alazar, N. (1990), "World oil and gas reserves and production forecasts," *Energy Exploration and Exploitation*, Vol.8.6, pp.380–92.

Baird, E. (2003), "Fossil fuels: the key to sustainable development," *World Energy*, Vol.6, No.1, pp.34–61.

Campbell, C.J. (1997), The Coming Oil Crisis, Brentwood, Multi-Science Publishing Co. Ltd.

Campbell, C.J. and Laherrère, J.H. (1998), "The end of cheap oil," *Scientific American*, March, pp.78–83.

Clawson, M. (Ed.) (1964), *National Resources and International Development*, Baltimore, Johns Hopkins University Press.

Econ Centre for Economic Analysis (1993), *Oil and Gas – a Sunset Industry?*, Oslo, Econ Centre.

Edwards, J.D. (1997), "Crude oil and alternative energy production forecasts for the 21st century," *AAPG Bulletin*, No.8, pp.292–305.

Fisher, J.C. (1974), *Energy Crises in Perspective*, New York, J. Wiley & Sons Ltd.

Gordon, R.L. (1987), *World Coal: Economics, Policies and Prospects*, Cambridge, Mass, Cambridge University Press.

Hartshorn, J.E. (1993), *Oil Trade; Politics and Prospects*, Cambridge, Cambridge University Press.

Heinberg, R. (2003), *The Party's Over: Oil, War and the Fate of Industrialised Societies*, British Columbia, New Society Publishing.

International Energy Agency (2003b), *World Energy Investment Outlook*, Paris, OECD.

Laherrère, J.H. (2003), "Future of oil supplies", *Energy Exploration and Exploitation*, Vol.21, No.3, pp.227–267.

Lang, D.A. et al. (2000), "Non-conventional hydrocarbons production," *Proceedings of the 16th World Petroleum Congress*, Calgary, Vol.2, pp.210–259.

McCabe, P.J. (1998), "Energy resources: cornucopia or empty barrel," *AAPG Bulletin*, Vol.82.11, pp.110–34.

Norwegian Ministry of Petroleum and Energy (2002), *Report on the Norwegian Continental Shelf*, Oslo.

Odell, P.R. (1971), *Oil and World Power*, London, Penguin Books, 1st edition.

Odell, P.R. (1986), *Oil and World Power*, London, Penguin Books, 8th edition.

Odell, P.R. (1997a), "The global oil industry: the location of production," *Regional Studies*, Vol.31.3, pp.309–320.

Odell, P.R. (2001), *Oil and Gas: Crises and Controversies, 1961–2000*, Vol.1, Brentwood, Multi-Science Publishing.

Odell, P.R. (2002), *Oil and Gas: Crises and Controversies, 1961–2000*, Vol.2, Brentwood, Multi-Science Publishing.

Oil and Gas Journal (2003b), "Future energy supply," *Oil and Gas Journal*, Vol.101, Nos.27–32.

Rees, J. (1985), *Natural Resources: Allocation, Economics and Policy*, London, Methuen.

Schurr, S. and Netschert, B. (1977), *Energy in the American Economy, 1850–1975*, Baltimore, The Johns Hopkins University Press.

Smith, N.J. and Robinson, G.H. (1997), "Technology pushes reserves crunch date back," *Oil and Gas Journal*, Vol.95, April 7, pp.43–50.

Chapter 6: Oil and Gas as Renewable Resources?

Biogenic Carbon Energies' Limitations

Chapters 2 to 4 have presented reasoned hypotheses for the evolution by source of global carbon energy supplies in the 21st century. This has shown that renewables will do no more than modestly supplement the increasingly important contributions of coal, oil and natural gas as they maintain the 2% per annum rate of growth in energy demand until mid-century. Thereafter, there will be a slowly declining rate of increase in the demand for energy. This will be down to only 1.2% per annum in the 2090s. As a consequence of this falling rate of increase in energy demand, the supply of renewables will still need to increase only modestly for much of the rest of the century. Indeed, any significant increases in the requirements for renewable energy supplies will, even in the later decades of the century, only occur if they are competitive with carbon fuels.

The 21st century required supply of coal will, as shown in Chapter 2, involve a cumulative total output of some 460Gtoe. This is much the same as the present declaration of proven reserves of 465Gtoe which, in turn, encompass only 8% of estimated ultimately recoverable coal resources. As it is near certain that additional reserves of coal, more than adequate to sustain the continuation of the relatively modest expansion of production from 2.2Gtoe in 2000 to 6.4Gtoe in 2100, will be declared proven in the 21st century, then a sufficiency of supplies to meet the demand is self-evident; especially in the context of the long-term increases in energy prices (in real terms) as hypothesised in Chapter 5. There is thus no requirement for any re-evaluation of the validity of the organic theory of coal's origins and its occurrence. The limitations on the contribution of coal to the energy supply in the 21st century to less than the level indicated in this study will be essentially environmental *per se* and/or the result of the failure to develop economically viable 'clean coal' technology and/or the large scale sequestration of CO_2 (Williams, 1998).

For oil and gas our assessments of conventional and non-conventional reserves and resources indicate that their joint production will continue to expand while they remain competitive with alternative sources. They can also be considered increasingly environmentally friendly in the context of the now ongoing development of cleaner producing and consuming technologies (Williams, *ibid*.). This will remain the situation until the 2060s when we have hypothesised a constraint arising from possible resources limitations. Although such a possible shortfall in their supply emerges from an overly pessimistic assessment of their ultimately recoverable supplies, it does, nevertheless, suggest an element of fragility in the long-term future during which meeting the demand for energy remains significantly dependent on the exploitation of hypothesised finite volumes of resources of oil and gas. Note, however, that the long-term future (viz. 70 years) for the potential shortfall is such as to allow plenty of time and scope for an intervening fundamental reappraisal of oil and gas supply-side limitations, both globally and regionally.

The Abiogenic Theory of the Origins of Hydrocarbons
The element of doubt, however, over the future availability of oil plus gas suggests that this appraisal of the probabilities of the adequacy of the world's oil and gas supplies should not exclude consideration of the validity of the original 18th century hypothesis that oil and gas are generated from biological matter in the chemical and thermodynamic environments of the earth's crust (Abbas, 1996). This is reasonable given that there is an alternative "modern Russian-Ukrainian" theory of the origins of oil and gas, even though it is one which generally seems to be treated with scepticism (Kenney, 1996) – or even scorn – in the West. This is reflected in the apparent absence of a willingness to publish reasoned critiques of the alternative theory. For example, in *The Coming Oil Crisis* (1997), the petroleum geologist, C. J. Campbell writes: 'few people take the hypothesis of an inorganic origin of oil seriously'. He then, however, demonstrates his apparent lack of familiarity with the subject by citing only one reference. Moreover, he designates that reference as the original paper on the theory, although it was not published until 1994 as a paper on a specific application of the then already 50-year-old inorganic theory of the origins of

petroleum to recent successes in drilling and developing oil and gas fields in pre-Cambrian rocks in the Ukraine (Krayuskin et al, 1994). A decade earlier, in a standard text on *Petroleum Geology* some pages were devoted to theories of the inorganic origins of oil and gas. This concluded that "our present stock of petroleum hydrocarbons would (given the abiogenic theory) represent biogenic additions to a fundamentally primordial endowment" (North, 1988).

Recent applications of the inorganic theory of the origin of petroleum have, however, led to claims for the possibility of the Middle East fields being able to produce oil 'forever' (Mahmoud and Beck, 1995) and to the concept of *repleting* oil and gas fields in the Gulf of Mexico, so that hydrocarbons can be re-defined as a 'renewable resource, rather than a finite one' (Gurney, 1997). There has also been the discovery of 12 fields on the flanks of the Dnieper-Donetz basin in the Ukraine, with recoverable oil reserves of 1.6 billion barrels and over 100 billion cubic metres of gas reserves, the major part of which 'is produced from the pre-Cambrian crystalline basement' (Krayushkin et al, 1994). Other indicated locations in the copious literature on abiogenic oil and gas occurrence include Algeria, China, India, Libya, the North Sea, the US (Kansas, Texas and Wyoming), Venezuela, Vietnam, Western Canada and, of course, widely in Russia (e.g. the north Caucasus, Komi, Siberia and Volga-Urals). More generally, it was already argued more than 25 years ago that 'all giant oil fields are most logically explained by inorganic theory' because 'simple calculation of potential hydrocarbon contents in sediments shows that organic materials are too few to supply the volumes of petroleum involved... the very fact of giant oil fields can refute the whole complex of argumentation in favour of organic theory' (Porfir'yev, 1974).

Given the scope and the complexities of the scientific evidence for and against the abiotic theory, it is difficult for a non-physical scientist to make a judgement on its validity or otherwise. It does not, however, seem to be any more inherently excludable for consideration as an explanation for the occurrence of hydrocarbons in the earth's crust than the organic hypothesis of the derivation of oil and gas. Its significance in an evaluation of the potentially available hydrocarbon resources for the 21st century is, however, self-evident. Instead of having to consider a stock reserve already accumulated in an unknown, but finite, number of so-called oil plays, there is the possibility of evaluating oil and gas as renewable reserves in the context of whatever demand development may emerge over an unlimited period. This is a quite fundamental issue in relation to the very long-term horizon of this study.

It raises quite different issues from the controversy over the prospects for the exploitation of non-conventional oil and gas, as presented in Chapters 3 and 4. These resources from alternative habitats have been presented as important volumetric additions to the availability of conventional hydrocarbons; but also as additions which become accessible for mankind's use only at generally higher costs. Such an economic restraint on non-conventional oil and gas production is quite distinct from the contrasting prospects for oil and gas as organically derived materials, on the one hand, and as abiotic occurrences, on the other. If fields do replete because the origin of the oil and gas extracted from them is abyssal and abiotic (based on chemical reactions under specific thermodynamic conditions deep in the earth's mantle), then extraction costs should not rise, as production can continue more or less undiminished for an indefinite period. Furthermore, estimates of oil and gas reserves, of reserves-to-production ratios and of annual rates of discovery and additions to reserves do not have any of the critical importance correctly attributed to them in evaluating the future prospects of supply in the context of the organic hypothesis of oil and gas' derivation. In essence, the stage on which consideration of the issues of the future availabilities of oil and gas has hitherto been played no longer remains appropriate.

If oil and gas were to be accepted as renewable resources by virtue of their inorganic derivation – in the same way that geothermal power has come to be accepted as a renewable, rather than a finite, resource within the past 30 years – they would establish a set of alternative carbon energy prospects for meeting global energy requirements in the 21st century, not only in scientific and technological terms, but also from the standpoints of economics and geopolitics.

Thus, the alternative theory of hydrocarbons' origin seems to be too important to be omitted from a study concerned with the very long-term prospects for energy supplies. Hence, this chapter continues with an edited extract from an article published by Dr J. F. Kenney of the Joint Institute of the Physics of the Earth, Russian Academy of Sciences, Moscow, and the president of the Gas Resources Corporation, Houston in *Energy World*, No. 240, June 1996. It is incorporated into this chapter with the author's permission.

"The modern Russian-Ukrainian theory of abyssal, abiotic petroleum origins is an extensive body of scientific knowledge covering the subjects of the chemical genesis of hydrocarbon molecules, the physical processes which occasion their terrestrial concentration, the dynamical processes of the movement of that

material into geological reservoirs of petroleum, and the location and economic production of petroleum. As stated, the modern theory has determined that petroleum is a primordial material of deep origin which is transported at high pressure via 'cold' eruptive processes into the crust of the Earth. The theory is almost unique among what too often pass as 'theories' in the field of geology (especially in the US) in that it is based not only upon extensive geological observation, but also upon rigorous, analytical, physical reasoning. Much of the modern Russian theory of abyssal, abiotic petroleum genesis developed from the sciences of chemistry and thermodynamics and accordingly the modern theory has steadfastly held, as a central tenet, that the generation of hydrocarbons must conform to the general laws of chemical thermodynamics – as must likewise all matter. With the exception of methane, the alkane of the lowest chemical potential of all hydrocarbons and, to a lesser extent, ethene, the alkene of the lowest chemical potential of its homologous molecular series, petroleum has no intrinsic association with biological material.

Methane is thermodynamically stable in the pressure and temperature regime of the near surface crust of the Earth and accordingly can be generated there spontaneously, as is indeed observed for phenomena such as swamp gas or sewer gas. However, methane is practically the sole hydrocarbon molecule possessing such characteristics in that thermodynamic regime. Almost all other reduced hydrocarbon molecules, excepting only the lightest ones, are high pressure polymorphs of the hydrogen-carbon system. The genesis of heavier hydrocarbons occurs only in multi-kilobar regimes of high pressures.

Because of the general lack of familiarity outside the former Soviet Union with the modern Russian-Ukrainian theory of abyssal, abiotic petroleum origins, several immediate facts about that body of knowledge deserve to be set forth;

– first, it is not new or recent. This theory was first enunciated by Professor Nikolai Kudryavtsev in 1951, over half a century ago, and has undergone extensive development, refinement and application since its introduction. There have been more than 4,000 articles published in the Soviet scientific journals and many books dealing with the modern theory.

– second, it is not the work of any one single man – nor of a few men. The modern theory was developed by hundreds of scientists in the former Soviet Union, including many of the finest *geologists, geochemists, geophysicists and thermodynamicists of that country. There have now been more than two generations of scientists who have worked upon, and contributed to, the development of the modern theory.*

– third, it is not untested or speculative. On the contrary, the modern theory was severely challenged by many traditionally-minded geologists at the time of its introduction. During the first decade thereafter, the modern theory was thoroughly examined, extensively reviewed, powerfully debated and rigorously tested. Every year following 1951 there were important scientific conferences organised in the Soviet Union to debate and evaluate the modern theory, its development and its predictions. The All-Union Conferences on Petroleum and Petroleum Geology in the years 1952–1964/65 dealt particularly with this subject. Indeed, during the period when the modern theory was being subjected to extensive critical challenge and testing, a number of the men who supported it pointed out that there had never been any similar critical review or testing of the traditional hypothesis that petroleum might somehow have evolved spontaneously from biological detritus.

– fourth, it is not a vague, qualitative hypothesis, but stands as a rigorous analytic theory within the mainstream of the modern physical sciences. In this respect, the modern theory differs fundamentally not only from the previous hypothesis of a biological origin of petroleum, but also from all traditional geological hypotheses. Since the 19th century, knowledgeable physicists, chemists, thermo-dynamicists and chemical engineers have regarded with grave reservations (if not outright disdain) the suggestion that highly reduced hydrocarbon molecules of high free enthalpy (the constituents of crude oil) might somehow evolve spontaneously from highly oxidised biogenic molecules of low free enthalpy. Beginning in 1964, Soviet scientists carried out extensive theoretical statistical thermodynamic analysis which established explicitly that the hypothesis of evolution of hydrocarbon molecules (except methane) from biogenic ones in the temperature and pressure regime of the Earth's near-surface crust was glaringly in violation of the second law of thermodynamics. They also determined that the evolution of reduced hydrocarbon molecules requires pressures of magnitudes encountered at depths equal to the mantle of the Earth. During the second phase of its development, the modern theory of petroleum was entirely recast from a qualitative argument based upon a synthesis of many qualitative facts into a

quantitative argument based upon the analytical arguments of quantum statistical mechanics and thermodynamic stability theory (Chekaliuk 1967; Boiko 1968. Chekaliuk 1971; Chekaliuk and Kenney 1991; Kenney 1995). With the transformation of the modern theory from a synthetic geology theory, arguing by persuasion, into an analytical physical theory, arguing by compulsion, petroleum geology entered the mainstream of modern science.

– fifth, it is not controversial nor presently a matter of academic debate. The period of debate about this extensive body of knowledge has been over for approximately two decades (Simakov 1986). The modern theory was applied extensively throughout the former Soviet Union as the guiding perspective for petroleum exploration and development projects. There are presently more than 80 oil and gas fields in the Caspian district alone which were explored and developed by applying the perspective of the modern theory and which produce from the crystalline basement rock. (Krayushkin, Chebanenko et al, 1994) Similarly, such exploration in the western Siberia cratonic-rift sedimentary basin has developed 90 petroleum fields of which 80 produce either partly or entirely from the crystalline basement. More recently, exploration discovered 11 major fields and one giant field on the northern flank of the Dnieper-Donetz basin. There are also deep drilling exploration projects under way in Azerbaijan, Tatarstan and Asian Siberia directed to testing potential oil and gas reservoirs in the crystalline basement.

The errors involved in predictions about the future availability of petroleum, inevitably occasioned by an inappropriate application of the rococo hypothesis that petroleum somehow miraculously evolved from limited volumes of biogenic matter, obtain generally from the very notion of oil as a limited 'fossil' material. Instead, one should better recognise that there exists no more reason to expect a future shortage of petroleum than of, say, mid-oceanic ridge basalt (MORB) which is the rock characteristic of the loci of the deep suture spreading zones on the mid-ocean floor, where new oceanic crust is constantly being erupted from the mantle of the Earth. Those predictive errors obtain specifically from the neglect of several extremely large potential sources of petroleum, viz.

1. the potential to produce petroleum from the crystalline basement, from volcanic structures, from impact structures and from non-sedimentary regions has generally been neglected.

2. the petroleum potential of the riftogenic suture zones, both subsea and on-shore, has been largely neglected.
3. the petroleum which certainly exists and will surely be produced from reservoirs underneath those presently being produced has been entirely neglected.
4. the potential to produce petroleum gas from reservoirs beneath the methane clathrate zones has been completely neglected, as have most of the methane clathrate reserves themselves.
5. the potential that certain of the petroleum fields presently producing may be drawing pressured hydrocarbons from an open and active fault or conduit from the mantle, and therefore, may never be depleted has been entirely neglected, as has the potential to develop non-depleting fields by deep drilling.

In view of these considerations, there stands no reason to worry about, and even less to plan for, any predicted demise of the petroleum industry based upon a vanishing of petroleum reserves. On the contrary, these considerations compel additional investment and development in the technology and skills of deep drilling, of deep seismic measurement and interpretation, of the reservoir properties of crystalline rock, and of the associated completion and production practices which should be applied in such non-traditional reservoirs.

Not only are any predictions that the world is 'running out of oil' invalid, so also are suggestions that the petroleum exploration and production industry is a 'mature' or 'declining' one. This writer's experience, gained from working in the former Soviet Union during the past decade, has provided compelling evidence that the petroleum industry is only now entering its adolescence."

The following references illustrate the issues set out above:

Anisimov, V. V., Vasilyev, V. G. et al (1959) 'Berezov gas-prone district, and perspectives of its development', *Geology of Oil and Gas*, 9: 1–6.

Boiko, G. E. (1968) *The transformation of abyssal petroleum under the conditions of the earth's crust*, Kiev, Naukova Dumka.

Chekaliuk, E. B. (1967) *Oil in the earth's upper mantle*, Kiev, Naukova Dumka.

Chekaliuk, E. B. (1971) *The thermodynamic basis for the theory of the abiotic genesis of petroleum*, Kiev, Naukova Dumka.

Chekaliuk, E. B. and Kenney J. F. (1991) 'The stability of hydrocarbons in the thermodynamic conditions of the Earth', *Proc. Am. Phys. Soc.* 36(3): 347.

Dolenko, G. E. (1968) 'The origin of oil and gas deposits in the crust of the Earth', *Geol. Zh.* 2:67.

Dolenko, G. E. (1971) *On the origin of petroleum deposits: the origin of petroleum and natural gas and the formation of the commercial deposits*, Kiev, Naukova Dumka, 3.

Kenney, J. F. (1995) 'The spontaneous high-pressure generation and stability of hydrocarbons: the generation of n-alkanes, benzene, toluene and xylene at multi-kilobar pressures', *Joint 15th AIR/APT International Conference on High-Pressure Physics and Technology*, Warsaw.

Krayushkin, V. A. (1965) *Theoretical problems of migration and accumulation of oil and natural gas: synopsis of theses for degrees of doctor of science*, Moscow, I. M. Gubkin Institute of the Oil, Chemical and Gas Industry, 36.

Krayushkin, V. A. (1984) *The abiotic, mantle origin of petroleum*, Kiev, Naukova Dumka.

Krayushkin, V. A., Chebanenko, I. I. et al (1994) 'Recent applications of the modern theory of abiogenic hydrocarbon origins: drilling and development of oil and gas fields in the Dnieper-Donetz basin', *Proceedings of the 7th International Symposium on the Observation of the Continental Crust through Drilling*, Santa Fé, New Mexico, DOSECC: 21–24.

Kropotkin, P. N. (Ed) (1956) 'Origin of hydrocarbons of the Earth's crust', *Proceedings of discussion on the problem of origin and migration of oil*, Kiev, Academy of Sciences Press, the Ukrainian SSR.

Kudryavtsev, N. A. (1951) 'On the problem of petroleum genesis and the formation of oil deposits', *Neft. Kh-vo.* 9:17–29.

Kudryavtsev, N. A. (1959) *Oil, gas and solid bitumens in igneous and metamorphic rocks*, Leningrad, State Fuel Technical Press.

Kudryavtsev, N. A. (1963) *Deep faults and oil deposits*, Leningrad, Gostoptekhizdat.

Letnikov, F. A., Karpov, I. K. et al (1977) *The fluid regime of earth crust and upper mantle*, Moscow, Nauka Press.

Linetskii, V. F. (1974) *The migration of oil and gas at great depths*, Kiev, Naukova Dumka.

Markevich, B. P. (1966) *The history of geological evolution and petroleum-content of the West Siberian lowland*, Moscow, Nauka Press.

Porfir'yev, V. B. (1959) *The problem of the migration of petroleum and the formation of accumulations of oil and gas*, Moscow, Gostoptekhizdat.

Porfir'yev, V. B. and Klochko, V. P. (1981) *Oil-content problem of basement of Siberia: geological and geochemical principles of prospect for oil and gas*, Kiev, Naukova Dumka Press: 36–101

Raznitsyn, V. A. (1963), 'Perspectives of petroleum-content of the Timan-Pechera region', *Petroleum Geology and Geophysics* 10: 27–31.

Simakov, S. N. (1986) *Forecasting and estimation of the petroleum-bearing subsurface at great depths*, Leningrad, Nedra.

The Theory Ignored to Date – Is this about to change?

Dr Kenney's brief but formidable exposure of the background to, and the main elements of, the scientific endeavours in the former Soviet Union that produced the abiogenic theory of oil and gas formation must, of course, be put in the political context of the Cold War between East and West. This produced a consequential near-absence of effective scientific contact for most of that time with only one major article on the subject by a main proponent of the theory appearing to have been published in a western journal dedicated to petroleum geology, viz. V. B. Porfir'yev's 'Inorganic origin of petroleum', published in the *Bulletin of the American Association of Petroleum Geologists* (Vol. 58, No.1, 1974, pp3–33). Even this, however, was preceded by an extraordinary and quite abnormal editorial caveat in which any form of responsibility by the Bulletin for the highly controversial views published was robustly disclaimed. There was no subsequent discussion on the paper.

The theory was thus, it seems, virtually shut out of the western world's consideration of future prospects for oil and gas occurrence. Even at the East-West scientific institution, specifically created in 1972 in Vienna following an earlier initiative by President Kennedy and Premier Khrushchev, viz. *The International Institute for Applied Systems Analysis* (IIASA), a major project on international energy prospects undertaken in the mid-to-late 1970s ignored a presentation by Academician Professor M. A. Styrikovich from Moscow indicating that 'a cautious view of ultimate world oil resources suggests an availability of 11,000 billion barrels' (Styrikovich, 1977). Instead, the IIASA study pessimistically concluded that there was a severe limit on future oil availability (Häfele, W. et al, 1978). The very much larger volumes predicted by Styrikovich did not, of course, fit in with western estimates based on oil's organic origins.

The still-emerging re-globalisation of science in the post-Soviet era is, however, now providing a window of opportunity for a thorough reappraisal of the two theories of the origin of oil and gas – in which the views of western 'oil-men' who are now being exposed to exploration and exploitation opportunities in the former Soviet Union, are particularly significant. An ultimate synthesis of the two theories may well emerge, as, indeed, recently suggested by the Nobel Prize winner, Dr Thomas Gold, in *The Deep Hot Biosphere* (1999). In this, Gold hypothesises an 8km or more subterranean microbiological habitat of immense proportions through which oil and gas of abiogenic origin from deep sources pass on their migration route up to near-surface reservoirs and, in so doing, secure biological markers of the types found in hydrocarbons, but hitherto thought to represent a residual from the original biological debris from which oil and gas have been created.

At the beginning of the 21st century, petroleum science thus still has a fundamental issue to resolve. From this all-important work, the long-term prospects for oil and gas could well be radically changed from the conclusion of this study, viz. that there are 'only' 70 years remaining to peak production, to one which indicates a continuing availability of oil and gas supplies.

Significantly, the first-ever American Association of Petroleum Geologists Hedberg Research Conference concerned with the controversy over the biogenic/abiogenic origins of oil and gas, will be held in Vienna in July 2004. Upwards of 80 papers have been submitted for presentation and discussion. The conference has been instigated by Dr. Michael Halbouty, the doyen of American petroleum geologists, in the light of his experience of the continuity of production from some of his company's oilfields in the Gulf of Mexico long after they should – by the norms of the industry – have been depleted. The Conference could well lead to a significant breakthrough to a better global understanding of the nature of oil and gas resources; and of their potential very long-term capabilities for energising the world's economies and societies, without fear of exhaustion.

References

(excluding those listed above in the extract from J.F. Kenney's paper.)

Abbas, S. (1996), "The non-organic theory of the genesis of petroleum." *Current Science*, Vol.71.9.

Campbell, C. (1997), *The Coming Oil Crisis*, Brentwood, Multi-Science Publishing Co. Ltd.

Gold, T. (1999), *The Deep Hot Biosphere*, New York, Copernicus SpringerVerlag.

Gurney, J. (1997), "Migration or replenishment in the Gulf," *Petroleum Review*, May, pp.200–3.

Häfele, W. et al (1978), *IIASA Project on Energy Systems*, Laxenburg, IIASA.

Halbouty, M.T. (2001), "Exploration in the new millennium," Tulsa, *AAPG Memoir*, No.74.

Kenney, J.F. (1996), "Impending shortage of petroleum re-evaluated." *Energy World*, No.240, June, pp.16–19.

Krayuskin V.A. et al (1994), "Recent applications of abiogenic hydrocarbons origins," *Proceedings of the 7th International Symposium on the Continental Crust*, Sante Fé, pp.21–4.

Mahmoud R.F. and Beck, J.N. (1995), "Why the Middle East fields may produce oil forever," *Offshore,* April, pp.56–62.

North, F. (1988), *Petroleum Geology*, Boston, Allen and Unwin.

Porfir'yev, V.B. (1974), "Inorganic origin of petroleum," *AAPG Bulletin*, Vol.58.1, pp.3–53.

Styrikovich, M.A. (1977), "The long-range energy perspective," *Natural Resources Forum*, Vol.1, No.3, pp.252–63.

Williams, R.H. (1998), "A technological strategy for making fossil fuels environment and climate friendly," *World Energy Council Review*, September, pp.59–67.

Postscript: February 2004

The first three years of the 21st century have already become history. Does the data for these years sustain the validity of trends forecast in the book? Have events and developments since 2000 indicated the onset of significant political or economic disturbances in the global energy system? This Postscript briefly examines the most important issues.

a. The demand for energy

Post-2000, the global demand for energy has continued to increase at the late-20th century rate of 1.7% per annum (see Fig. 1.1). The 3-year increase in use of some 425 mtoe has, however, been over 95% met by carbon fuels, given that hydro-electricity has failed to maintain late 20th century production levels, while nuclear power's expansion of over 2.5% per annum in the 1990s has virtually ground to a halt with the closure of time-expired nuclear power stations and little by way of new facilities.

Meanwhile, the expansion of new renewables (such as wind and biomass energy) has barely got into its stride, in spite of the large government subsidies available for such developments in most OECD countries.

In overall terms, therefore, global energy use since 2000 has become more, rather than less, dependent on carbon fuels. This is at variance with the forecast in Chapter 1 for a short-term future increase in the contribution of renewables to total energy demand as policies which aimed to encourage the production of clean energy, in the context of the very high profile fears for the adverse consequences of global warming/climate change and in line with commitments of most of the industrialised countries to CO_2 emissions' reductions, were introduced. These policy actions should, moreover, have been even more strongly motivated by the persistence of much higher carbon fuels' prices since 1999 as a consequence of the re-establishment of an "ordered" international oil market by the oil exporting countries, supported by the United States (Odell, 2001).

If even these highly favourable conditions for the expansion of renewable energies have failed to generate any increase to date in their contribution to total energy use, then the 2000–2020 trend suggested in this study for a temporary doubling of renewables share of the market to 20% (see Fig.1.4) now seems unlikely to be achieved. Instead, the "best" one can now expect is their very slow expansion through the first half of the century to an eventual 20% contribution to energy demand by 2050.

Meanwhile, the highly pro-active national policies (especially in Europe) which are intended to stimulate the production of electricity by non-CO_2 emitting direct and indirect solar energy developments, are being increasingly called into question. This is partly for aesthetic reasons – as, for example, in concerns over the "blots" on the landscape – or event the seascape – created by wind-farms. In greater part, however, it is because of economic concerns over their development. The always higher – and, generally, much higher – capital costs per unit of electricity capacity/output from non-carbon energy sources (IEA, 2003) necessitates the payment of subsidies to the investing companies – and, hence, increased taxes to generate the public funds needed; and/or higher prices for electricity consumers; and/or so-called environmental taxes on inherently lower cost carbon energy based electricity production.

Whichever way is used to secure the production of electricity from renewables – based on the objective of internalising all environmental costs – the end result is unacceptable to most of the general public. So much so, indeed, that the desired objectives of the policy makers become politically unacceptable and so threaten future elected governments. In essence, the potentially offsetting benefits of switching from carbon to renewable energy sources are so far in the future as to be of minimal interests to today's electorates. This effectively inhibits progress towards the use of so-called "sustainable" energy provision, even in the world's richer countries, let alone in the much larger number of poor countries.

b. Energy Resources and Supplies
Since 2000, the declared proven reserves and the estimated resources

of carbon fuels have continued to grow more than quickly enough to offset the rate of depletion generated by the modestly rising demands for energy. Over the past three years some 76 billion barrels of conventional oil (10 Gtoe) have been produced, but, notwithstanding, remaining declared proven reserves have increased by almost 60 billion barrels (8 Gtoe). This means that gross reserves additions (from new discoveries and reserves' growth in existing fields) have been 136 billion barrels (18 Gtoe); to give a ratio of 1.8 between reserves additions and reserves depleted. Meanwhile, the declared proven reserves of non-conventional oil which were close to zero in 2000 are now defined at over 175 billion barrels (24 Gtoe). Overall, the global production to reserves ratio for oil now stands at a record all-time high of 45 years.

Meanwhile, a new evaluation (Linden, 2004) of the world's remaining ultimately recoverable oil reserves and resources of conventional plus non-conventional oil indicates a volume of 8000 billion barrels (1090 Gtoe). This may be compared with the much more modest 5000 billion barrels of remaining recoverable resources which have been used in this study: on the basis of which global oil production is shown as reaching its peak 49 billion barrels (6.7 Gtoe) in 2060. Linden also calculates the peak as occurring in 2060, but at a significantly higher level of 65 billion barrels (8.9 Gtoe), given his expectation for a higher rate of demand growth.

For natural gas the post-2000 growth in the knowledge of its occurrence and of the technology required for its exploitation has been even more emphatic than for oil (Adelman and Lynch, 2003). From 2000 to 2003 the declared proven reserves of conventional gas have increased from 135 Gtoe (150 Tcm) (see Table 4.1) to an end-2003 estimate of 155 Gtoe (172 Tcm); in spite of a production of almost 7 Gtoe (8 Tcm) over the three-year period. Thus, reserves' growth has outstripped use by a factor of three, to produce a current R/P ratio of almost 75 years. Such a strengthening of the reserves' base suggests that the curve for conventional gas production, as set out in Figs. 4.1 and 4.2, is well below that which could be achieved, were a higher growth in demand to develop.

Linden (2004) anticipates a future for gas – including some non-conventional gas in the United States – which reflects a volume of resources four times higher than the presently declared proven reserves. This is sufficient gas to sustain production growth to a peak of 6.5 Gtoe (7.23 Tcm). This compares with this book's forecast for a 5.4 Gtoe (6.1 Tcm) peak production from a remaining resource base of conventional gas of about 415 Tcm (375 Gtoe). This study, however, goes on to hypothesise (see pp.76–9) the depletion of an additional 305 Gtoe (340 Tcm) of non-conventional gas over the period from 2020–2100 (Figs. 4.1 and 4.2), whereby the global gas industry continues to grow until 2090 with a peak production in that year of 11.2 Gtoe (12.4 Tcm). Linden's study does not consider this potential for further development based on estimated global non-conventional gas resources, so producing a requirement for additional alternative energies – notably coal and nuclear power – in the second half of the 21st century. The contrast between these forecasts necessitate further examination: as does Linden's omission of any consideration of an eventual supply from gas hydrates (cf. pp.70–80 in this study). Heavily financed studies on the latter have recently been initiated in Japan and the United States (Rach, 2004).

c. Emerging Geopolitical "Big-issues"
i. An "ordered" international oil market
Moves to consolidate the tentative late-1990s steps towards the re-creation of such a market, through a US initiative supported by Russia and accepted by OPEC, with guarantees for high and stable crude oil prices, are under way. It has, indeed, just been institutionalised by the establishment of a permanent base and secretariat in Saudi Arabia for the IEA/OPEC-created International Energy Forum. This, in the words of the recently appointed Norwegian Secretary General of the Forum, has "the responsibility for facilitating a global energy dialogue between 170 countries." (MEES, 2004) If this significant geopolitical development is successful, then international oil and gas will, in essence, no longer be treated as "just another commodity."

ii. Russian/other CIS members' emerging solidarity
One now sees a rapidly emerging Russian/CIS objective to enhance the role of the hydrocarbons'–rich, former member states of the Soviet Union in the international oil and gas systems. The decline in Russia and the other republics' share of global oil and gas production in the post-1991 aftermath of the break-up of the Soviet Union has, over the past three years been largely restored, viz. to 20% of total global output. Meanwhile, their internal energy use remains 25% down from 1992. As a consequence of these developments Russia and the other CIS countries are now established as the world's most important hydrocarbons' exporter. Though difficulties remain between Russia and the other states (particularly over exports' infrastructure), there is

an emerging over-arching communality of interests between them, vis a vis other relevant main actors on the global energy scene – notably Europe, as the main importer of Russian/CIS oil and gas (Gas Matters, 2004), but also the rival producers/exporters of oil and gas in the Middle East.

iii. China as a world market leader for oil and gas imports
China is achieving a higher profile in global energy markets by virtue of its burgeoning demands for energy as a consequence of its rapid economic development. Since 2000 its energy use has grown by 25%, to make it the world's second largest consuming country (after the United States). Though it remains a small net exporter of coal and continues to expand its indigenous oil and gas production, it is already the fourth largest oil importer (after the US, Japan and – for only one more year – Germany). Its annual imports will continue to grow, while natural gas imports (both as LNG and pipelined gas) will start shortly and then expand rapidly as the required infrastructure for transmission and use is put-in-place. Thus, over the next two decades China will become a major player in international energy trade: and also an important investor in upstream oil and gas operations across Asia – including Russian Siberia, the Central Asian republics and, in particular, the Middle East – and in Australasia, where it has already purchased a large equity interest in gas production and its processing for export as LNG.

iv. The intensification of US concerns for access to oil and gas
The US' long-standing interests in and influence over, international oil and gas affairs (see section i. above) will be maintained for geopolitical reasons. (Maugeri, 2003) To this hegemonic function, however, it must also act internationally, when and where appropriate, to limit the perceived adverse impact of its growing dependence on energy imports; as its demand for *additional* oil and gas first approaches, and then surpasses, 200 mtoe per decade. As noted above in the case of China, this will involve the continuation of US companies' interests in oil/gas exporting countries. The two countries' thus seem likely to become serious rivals for establishing control over the world's hydrocarbons resources.

v. And how will it be for the Middle East?
It will certainly retain its role as the primary source of oil for export. It will even achieve a significant, if not a leading, role as a source of

natural gas for export; both as LNG for remote markets – such as the US and China – and by pipelines to serve the European and Indian sub-continents. These developments will help ensure the continuity of "cauldron" politics in the region. Remaining open to question, however, is *how* its role as the hydrocarbons supplier of last resort to so much of the rest of the world will be maintained. Perhaps the solution to this will emerge through the achievement of successful dialogue generated by the International Energy Forum (see c.i. above): or, if not, then by competing and/or conflicting American, Russian and Chinese interventions in the region – with each mega-power seeking a guarantee for its own specific interests.

References
Adelman, M.A. and Lynch, M.C. (2003), *Natural Gas Supply to 2100*, Hoersholm, International Gas Union.
*Anon. (2003), "A single Eurasian gas market; hope or hype?", *Gas Matters*, December 2003, pp.36–40.
International Energy Agency (2003b), *World Energy Investment Outlook*, Paris, OECD/IEA.
*Linden, H.R. (2004), "Rising expectations of ultimate oil and gas recovery to have critical impact on energy and environment policy", Part 1, *Oil and Gas Journal*, Vol.102.3, January 19, pp.18–28: Part 2, *Oil and Gas Journal*, Vol.102.4, January 26, pp.18–27.
*Maugeri, L. (2003), "Time to debunk mythical links between oil and politics", *Oil and Gas Journal*, Vol.101.48, December 15, pp.18–28.
*Middle East Economic Survey (2004), "Interview with the Secretary General of the International Energy Forum", Vol.47.5, February 2.
Odell, P.R. (2001), *Oil and Gas – Crises and Controversies, 1961–2000*, Vol.1, Global Issues, pp.456–54
*Rach, N.M. (2004), "Japan undertakes ambitious hydrate drilling program", *Oil and Gas Journal*, Vol.102.6, pp.37–9.

*These entries are not listed in the Bibliography

Bibliography
(Including references listed at the end of each chapter.)

Aalund, L.R. (1998), "Technology and money unlocking vast Orinoco reserves," *Oil and Gas Journal*, Vol.96, October 19, pp.49–72.

Abbas, S. (1996), "The non-organic theory of the genesis of petroleum," *Current Science*, Vol.71.9.

Adelman, M.A. (1972), *The World Petroleum Market*, Baltimore, The Johns Hopkins University Press.

Adelman, M.A. (1993), *The Economics of Petroleum Supply*, Cambridge, Mass., MIT Press.

Adelman, M. A. (1995) *The Genie Out of the Bottle: World Oil Since 1970* Cambridge, Mass., MIT Press.

Adelman, M. A. and Lynch, M. C. (1997) 'Fixed view of resource limits creates undue pessimism,' *Oil and Gas Journal, Vol.95,* 7 April, p.56

Adelman, M.A. and Lynch, M.C. (2003), *Natural Gas Supply to 2100*, Hoersholm, International Gas Union.

Alazard, N (1990), "World oil and gas reserves and production forecasts," *Energy Exploration and Exploitation*, Vol.8.6, pp.380–92.

Alazard, N. (1996) 'Technical and scientific progress in petroleum exploration/ production, impact on reserves and costs,' *Energy Exploration and Exploitation* Vol. 14, No. 2, pp.103–118.

Anderson, D. (2001), "Energy and economic prosperity," *World Energy Assessment*, New York. United Nations Development Programme.

Anderson, R. N. (1998) 'Oil production in the 21st century,' *Scientific American*, March, pp.86–91.

Anthrop, D.E. (2003), "Hydrogen and Fuel cells," *Oil and Gas Journal*, Vol.101.39, pp.10–15.

Apport, O. (2003), "The contribution of Technology; 'creating' reserves", MEES, Vol.LXVII, No.3, pp.D1–6.

B.P. (1979), *Oil Crisis… Again*, London, BP Policy Review Unit.

Badakhshan, A. (1997) 'Natural gas and its share of present and future global energy supply and demand,' *Proceedings of the 15th World Petroleum Congress*, Beijing.

Baird, E. (2003), "Fossil fuels: the key to sustainable development," *World Energy*, Vol.6, No.1, pp.34–61.

Baktiari, A.M. (2003), "Russia's gas production and exports," *Oil and Gas Journal*, Vol.101.10, pp.20–31.

Barnes, P. (1993), *The Oil Supply Mountain: is the Summit in Sight?* Oxford, Oxford Institute for Energy Studies.

Bartsch, U. and Müller, B. (2000), *Fossil Fuels in a Changing Climate: Impacts of the Kyoto Protocol and Developing Country Participation*, Oxford University Press for Oxford Institute for Energy Studies.

Beck, P. (1994), *Prospects and Strategies for Nuclear Power*, London, Royal Institute of International Affairs.

Bossel, H. (1998), *Earth at a Crossroads; Paths to a Sustainable Future*, Cambridge, Cambridge University Press.

Bradley, R.L. (2003), *Climate Alarmism Reconsidered*, London, Institute of Economic Affairs.

Brookes, L. (2000), "Energy Efficiency Fallacies Revisited", *Energy Policy*, Vol.28.4, pp349–60.

Campbell, C.J. (1997), *The Coming Oil Crisis*, Brentwood, Multi-Science Publishing Co. Ltd.

Campbell, C. J. (1998), 'Running out of gas: this time the wolf is coming,' *The National Interest*, Spring, pp.47–55.

Campbell, C.J. (2003), *The Essence of Oil and Gas Depletion*, Brentwood, Multi-Science Publishing Co.

Campbell, C.J. and Laherrère, J.H. (1998), "The end of cheap oil," *Scientific American*, March, pp.78–83.

Campbell, R.W. (1976), *Trends in the Soviet Oil and Gas Industry*, Baltimore, Johns Hopkins University Press.

Canadian Gas Potential Committee (2001), *Natural Gas Potential in Canada*, Calgary.

Carroll, J. (2003), *Natural Gas Hydrates*, Burlington, Mass; Gulf Professional Publishing.

Centre for Global Energy Studies (2001), *Oil Potential in the Middle East*, London.

Chandra, A. (2003), "Sizeable finds from India's licensing efforts," *Oil and Gas Journal*, Vol.101.23, pp.41–4.

Cherkashav, G.A. and Solawiev, V.A. (2002), "Economic use of hydrates: dream or reality," *Proceedings of the 17th World Petroleum*

Congress, Vol.1, pp.253–62.

Chow, L. (Ed.) (2003), "Themes in current Asian energy," *Energy Policy Special Issue*, Vol.31, No.11.

Ciriacy-Wantrup, S.V. (1952), *Resource Conservation, Economics and Policies*, Berkeley, University of California Press.

Claes, D.H. (2001), *The Politics of Oil Producer Cooperation*, Boulder, Westview Press.

Clawson, M. (Ed.) (1964), *National Resources and International Development*, Baltimore, Johns Hopkins University Press.

Colitti, M. and Simeoni, C. (1996) *Perspectives of Oil and Gas: the Road to Interdependence*, Dordrecht, Kluwer Academic Publishers.

Considine, J.I. and Kerr, W.A. (2002), *The Russian Oil Economy*, Cheltenham, Elgar Publishing.

Cornot-Gandolphe, S. (1995), "Changes in world natural gas reserves and resources." *Energy Exploration and Exploitation*, Vol.13.1, pp.3–18.

Correljé, A. et al (2003), *Natural Gas in the Netherlands*, Amsterdam, Oranje-Nassau Group.

Dafter, R. (1985), *Winning More Oil*, London, Financial Times Business Information.

Darmstadter, T. et al. (1977), *How Industrial Societies use Energy*, Baltimore, Johns Hopkins University Press.

De'Ath, N. G. (1997), 'Risk analysis in international exploration: a qualitative approach'; *World Oil*, September, pp.63–66.

Delahaye, C. and Grenon, M. (Eds.) (1983), *Conventional and Unconventional World Natural Gas Resources*, Laxenburg, IIASA.

Deffeyes, K.S. (2001), *Hubbert's Peak*, Princeton, Princeton University Press.

Desai, H. et al. (Eds.) (1987), Special Lesser Developing Countries Issue, *Energy Journal*, Vol.8.

Dienes, L. and Shabad, T. (1979), *The Soviet Energy System*, Washington D.C., Winston Press.

Dimitrevsky, A.N. (2002), "Environmental problems in developing oil and gas reserves of Russian Arctic areas," *Proceedings of the 17th World Petroleum Congress*, Vol.2, pp.551–60.

Douard, A. (2002), "Oil, gas, hydrogen and electricity; energies of the future for transport," *Proceedings of the 17th World Petroleum Congress*, Rio de Janeiro, Vol.1, pp.303–18.

Downey, M.W. et al. (2001), *Petroleum Provinces of the 21st Century*, Tulsa, Oklahoma, AAPG Memoir 74.

Dunkerley, J. (1981), *Energy Strategies for Developing Nations*, Baltimore, Johns Hopkins University Press.

Econ Centre for Economic Analysis (1993), *Oil and Gas – a Sunset Industry?* Oslo, Econ Centre.

Edge, G. (2003), "European renewable targets in danger," *EU Energy*, Vol.69, October 24, pp.22–4.

Edwards, J.D. (1997), "Crude oil and alternative energy production forecasts for the 21st century: the end of the hydrocarbon era," *AAPG Bulletin*, No.8, pp.292–305.

El Hinnawi, E.E. (1981), *The Environmental Impacts of Production and Use of Energy*, London, Tycooly Press.

Ellis, J. and Tréanton, K. (1998), 'Recent trends in energy-related CO_2 emissions,' *Energy Policy* Vol. 26, No. 12, February, pp159–66.

Energy Information Agency (1994a), *Energy Use and Carbon Emissions: some International Comparisons*, Washington, DC, US Government Printing Office.

Energy Information Agency (1994b), *Energy Use and Carbon Emissions; non-OECD Countries*, Washington DC, US Government Printing Office.

Energy Information Agency (2000), *Long-term World Oil Supply*, Washington DC, US Government Printing Office.

Estrada, J. et al (1995), *The Development of European Gas markets: Environmental Economic and Political Perspectives*, Chichester, J. Wiley & Sons Ltd.

European Commission (2001), *Towards a European Strategy for the Security of Energy Supply,* Brussels.

European Commission (2003), *Towards a Hydrogen-based Energy Economy,* Brussels.

European Environment Agency (2002), *Energy and Environment in the European Union*, Brussels.

Faithful, T.W. (2002), "Principles to practice: striving for sustainable development in an energy megaproject", *Proceedings of the 17th World Petroleum Congress*, Rio de Janeiro, Vol.5, pp.77–84.

Fisher, J.C. (1974), *Energy Crises in Perspective*, New York, J. Wiley & Sons Ltd.

Flowers, A. and King, R. (2002), *LNG today: the Promise and the Pitfalls*, Oxford, Energy Publishing Network.

Freund, P. (2002), "Technology for avoiding CO_2 emissions." *Proceeding of the 17th World Petroleum Congress*, Rio de Janeiro, Vol.5. pp.11–21.

Fritsch, B. (1982), *The Energy Demand of Industrialised and Developing*

Countries, Zurich, Institute of Technology.

Gas Matters (2001), "World gas majors seek involvement in cross-China pipeline project." London, August.

Gas Matters (2002), "Iran rekindles European gas export ambitions." London, February, pp.11–15.

Gas Matters (2003a), "Coal-bed methane creeps up the alternatives chart." London, August, pp.15–21.

Gas Matters (2003b), "Can India show enough market potential to lure new gas supplies," London, August, pp.3–10.

Gerholm, T.R. (Ed.) (1999), *Climate Policy after Kyoto*, Brentwood, Multi-Science Publishing.

Gires, J-M. (2000), "The Orinoco belt: new frontiers for long-term global reserves," *Proceedings of the 17th World Petroleum Congress*, Calgary, Vol.2, p.171.

Gold, T. (1979), "Terrestrial sources of carbon and earthquake outgassing," *Journal of Petroleum Geology*, Vol.1, No.3, pp.3–19.

Gold, T. (1999), *The Deep Hot Biosphere*, New York, Copernicus SpringerVerlag.

Gold, T. and Soter, S. (1982), 'Abiogenic methane and the origin of petroleum,' *Energy Exploration and Exploitation*, Vol. 1, No. 2, pp.89–102.

Goldenberg, J. and Coelho, S.T. (2004), "Renewable Energy: traditional v. modern biomass", *Energy Policy*, Vol.32.6, pp.711–4.

Gordon, R.L. (1987), *World Coal: Economics, Policies and Prospects*, Cambridge, Mass., Cambridge University Press.

Green, R.P. and Gallagher, J.M. (Eds.) (1980), *Future Coal Prospects: Country and Regional Assessments*, Cambridge, Mass., Ballinger.

Greenpeace (1997), *Putting a Lid on Fossil Fuels*, London, Greenpeace, UK.

Gregory, K. and Rogner, H-H. (1998), *Energy Resources and Conversion Technologies for the 21st Century*, Vienna, UN International Atomic Energy Agency.

Grenon, M. (1979), *Methods and Models for assessing Energy Resources*, Oxford, Pergamon Press.

Griffiths, J. (2001), "Fuel cells – the way ahead." *World Petroleum Congress Report*, London, I.S.C. Ltd.

Grimston, M.C. and Beck, P. (2002), *Double or Quits: the Global Future of Nuclear Power*, London, Royal Institute of International Affairs.

Groenveld, M.J. et al (2002), "Will the carbon age terminate before the depletion of resources?" *Proceedings of the 17th World Petroleum Congress*, Rio de Janeiro, Vol.1, pp.133–47.

Grossling, B.F. (1981), *World Coal Resources*, 2nd Edition, London, Financial Times Business Information.

Grossling, B. F. and Nielsen, D. J. (1985), *In Search of Oil*, Vol. 1, *The Search for Oil and its Impediments*, London, Financial Times Business Information.

Grossling, B. F. and Nielsen, D. J. (1985), *In Search of Oil*, Vol. 2, *Country Analyses*, London, Financial Times Business Information.

Grübler, A. et al. (1999), "Dynamics of energy technologies and global change," *Energy Policy*, Vol.27, pp.247–280.

Guilmot, J.F. (1986), *Energy 2000*, Cambridge, Cambridge University Press.

Gürer, N. and Bon, J. (1997), 'Factors affecting energy-related CO_2 emissions; past levels and present trends,' *OPEC Review* Vol. 21, No. 4, pp309–350.

Gurney, J. (1997), "Migration or replenishment in the Gulf (of Mexico)." *Petroleum Review*, May. pp.200–3.

Häfele, W. et al. (1978), *IIASA Project on Energy Systems*, Laxenburg, IIASA.

Haites, E.F. and Rose, A. (1996), "Energy and greenhouse gas mitigation." *Energy Policy Special Issue*, Vol.24.10/11, pp.857–1016.

Halbouty, M.T. (2001), "Exploration in the new millennium," in Downey, M.W. et al, *Petroleum Provinces of the 21st Century*, Tulsa, AAPG Memoir 74.

Hall, C. et al. (2003), "Hydrocarbons and the evolution of human culture," *Nature*, Vol.426, No.6964, pp.318–322.

Hancher, L. (2004), "The ins and outs of energy subsidies", *Platts EU Energy*, No.74, pp.22–3.

Harper, F.G. (1999), *Ultimate Hydrocarbon Resources in the 21st Century*, Birmingham, Alabama, AAPG.

Haas, A. et al. (2004), "How to promote renewable energy systems successfully and effectively", *Energy Policy*, Vol.32.6, pp.833–9.

Hartshorn, J.E. (1993), *Oil Trade; Politics and Prospects*, Cambridge, Cambridge University Press.

Hatcher, D. B. and Tussing, A. R. (1997), 'Long reserves lives sustain prospects for independents in the US Lower 48,' *Oil and Gas Journal*, Vol.49.46, pp.49–59

Head, I. (2003), "Biological action in the deep sub-surface and the origin of heavy oil," *Nature*, Vol.426, No.6964, pp.344–52.

Heinberg, R. (2003), *The Party's Over; Oil, War and the Fate of Industrialized Societies*, British Columbia, New Society Publishing.

Hiller, K. (1997), "Future world oil supplies – possibilities and constraints," *Energy Exploration and Exploitation*, Vol.15.2, pp.127–36.

Hoffman, G.W. (1985), *The European Energy Challenge; East and West*, Durham, North Carolina, Duke University Press.

Hoffman, P. (2001), *Tomorrow's Energy; Hydrogen, Fuel Cells and the Prospect for a Cleaner Planet*, Cambridge, Mass., Cambridge University Press.

Hoffman, J. and Johnson, B. (1981), *The World Energy Triangle*, Cambridge, Ballinger Press.

Holmes, C. (2003), "Uncertainties for FSU projects threaten export potential," *Oil and Gas Journal*, Vol.101.22.

Hols, A. (1972), "Future energy supplies to the Free World," *E.I.U. International Oil Symposium*, London, pp.1–24.

Holtberg, P. and Hirch, R. (2003), "Can we identify limits to worldwide energy resources?" *Oil and Gas Journal*, Vol.101.25, pp.20–6.

Horsnell, P. (1997), *Oil in Asia*, Oxford, Oxford University Press.

Hunter, H. I. (Ed.) (1964), *Introduction to World Resources*, New York, Harper and Row.

Huntingdon, H.G. and Brown, S.P.A. (2004), "Energy security and global climate change mitigation", *Energy Policy*, Vol.32.6, pp.715–8.

I.E.A. (1984), *World Energy Outlook*, Paris, OECD.

I.E.A. (1997), *Energy Technologies for the 21st Century*, Paris, OECD.

I.E.A. (1999), *CO_2 Emissions from Fuel Combustion*, Paris, OECD.

I.E.A. (2002a), *Flexibility in Natural Gas Supply and Demand*, Paris, OECD.

I.E.A. (2002b), *Developing China's Natural Gas market*, Paris, OECD.

I.E.A. (2002c), *World Energy Outlook to 2030*, Paris, OECD.

I.E.A. (2003), *Energy to 2050; Scenarios for a Sustainable Future*, Paris, OECD/IEA.

I.E.A. (2003a), *South American Gas: Daring to Tap the Bounty*, Paris, OECD.

I.E.A. (2003b), *World Energy Investment Outlook*, Paris, OECD/IEA.

I.I.A.S.A. (1995), *Global Energy Perspectives to 2050 and Beyond*, London, World Energy Council.

International Gas Union (1997), "World gas prospects: strategies and economics," *20th World Gas Conference*, Copenhagen.

International Gas Union (2003), "Catalysing an eco-responsible future," *21st World Gas Conference*, Tokyo.

I.P.C.C. (2001), *Climate Change*, Cambridge, Cambridge University Press.

Ismail, I. (1994), 'Future growth in OPEC oil production capacity and the impact of environmental measures,' *Energy Exploration and Exploitation*, Vol. 12, No. 1. pp17–58.

Ivanhoe, L. F. (1996), 'Updated Hubbert curves analyse world oil supply,' *World Oil* Vol. 217, No. 11, pp.91–94.

Jansen, J. C. et al. (1995), *Long Term Prices of Fossil Fuels*, Energie Centrum Nederland, Petten.

Jean-Baptiste, P. and Ducroux, R. (2003), "Energy policy and climate change," *Energy Policy*, Vol.31.2, pp.155–166.

Jenkins, G. (1997), 'World oil reserves reporting, 1948–1996: political, economic and subjective influences,' *OPEC Review*, June, pp.89–110.

Jensen, J.T. (2003), "The LNG revolution," *The Energy Journal*, Vol.24, No.2, pp.1–46.

Jones, D. (1998) *Climate after Kyoto – Outcome and Implications for Energy*, London, Royal Institute of International Affairs.

Johnston, D. (2003), *International Exploration Economics, Risk and Contract Analysis*, Tulsa, Oklahoma, Penn Well Corporation.

Kennedy, C. (2003), "The expansion of Russia's Siberian export capacity," *Oxford Energy Forum*, August, pp.12–17.

Kenney, J.F. (1996), "Impending shortage of petroleum re-evaluated," *Energy World*, No.240, June, pp.16–19.

Kalyuzhnova, Y. et al. (Eds.), (2002), *Energy in the Caspian basin*, Basingstoke, Palgrave.

Khartukov, E.M. (1997), "The control of Russia's oil," *Energy Exploration and Exploitation*, Vol.15.2, pp.117–26.

Khartukov, E.M. and Starostina, E. (2003), "Russia's new pipelines will de-bottleneck exports and production, " *Oil and Gas Journal*, Vol.101.38, pp.62–73.

Klare, M. (2002), *Resource Wars: the New Landscape of Global Conflict*, New York, Metropolitan Press.

Kleveman, L. (2003), *The New Great Game: Blood and Oil in Central Asia*, London, Atlantic Books.

Koch, F. (2002), "Hydropower, society and the environment," *Energy Policy Special Issue*, Vol.30, No.14.

Kolk, A. and Levy, D. (2001), "Winds of change: corporate strategy, climate change and oil multinationals," *European Management Journal*, Vol.19, No.5, pp.501–9.

Krayuskin V.A. et al. (1994), "Recent applications of abiogenic hydrocarbons origins." *Proceedings of the 7th International Symposium on the Continental Crust*, Sante Fé, pp.21–4.

Kruuskraa, V. A. et al. (1998), 'Emerging US gas resources,' *Oil and Gas Journal,* Issues from 13 April and 8 June.

Krylov, N.A. et al. (1997), "Exploration concepts for the next century," *Proceedings of the 15th World Petroleum Congress*, Beijing.

Kurtz, D. (1997), *Natural Gas in Latin America: Development and Privatisation*, London, Financial Times Business.

Labohm, H. et al., (2004), *Man-Made Global Warming: Unravelling a Dogma,* Brentwood, Multi-Science Publishing Co. Ltd.

Laherrère, J.H. (1997), "Production, decline and peak reveal true reserves figures." *World Oil*, pp.77–83.

Laherrère, J.H. (1998), "Development ratio evolves as true measure of exploitation." *World Oil*, February, pp.117–20.

Laherrère, J.H. (2003), "Future of oil supplies." *Energy Exploration and Exploitation*, Vol.21, No.3, pp.227–267.

Lainier, D. (1998), *Heavy Oil: a Major Energy Source for the 21st Century*, Edmonton, UNITAR.

Lang, D.A. et al. (2000), "Non-conventional hydrocarbons production," *Proceedings of the 16th World Petroleum Congress*, Calgary, Vol.2, pp.210–259.

Lazarus, M. et al. (1993), *Towards A Fossil Free Energy Future: the Next Energy Transition,* Boston, Environment Institute.

Leach, G. et al. (1979), *A Low Energy Strategy for the United Kingdom,* London, International Institute for Economic Development.

Lenin, V. I. (1966). *Collected Works of V. I. Lenin,* Vol. 31, Moscow, Progress Publishers.

Lerche, I. (2000), "Gas hydrates," *Energy Exploration and Exploitation*, Vol.18, Nos.2 and 3.

Lerche, I. (Ed.). (2002), "Gas in the 21st Century," *Energy Exploration and Exploitation,* Vol.20, No.4 (a collection of papers on regional prospects).

Linden, H.R. (1998), "Flaws in resources' models behind forecasts for oil supply and price," *Oil and Gas Journal*, Vol.94, No.52, pp.33–7.

Linden, H.R. (2004), "Rising expectations of ultimate oil and gas recovery to have critical impact on energy, environmental policy", *Oil and Gas Journal*, Vol.102.3, pp.18–28, (Part 1) and Vol.102.4, pp.18–28, (Part 2).

Lowrie, A. and Max, M.D. (1999), "The extraordinary promise and challenge of gas hydrates." *World Oil*, September, pp.49–55.

Lynch, M.C. (1997), *The Wolf at the Door or Crying wolf: Fears about the next Oil Crisis,* Cambridge, Mass., Centre for International Studies, M.I.T.

Mabro, R. and Wybrew-Bond, I. (Eds.) (1999), *Gas to Europe: the Strategies of the Major Suppliers*, Oxford, Oxford University Press.

MacKenzie, J. J. (1996), *Oil as a finite resource: when is global production likely to peak?*, World Resources Institute, Washington DC.

Mahmoud R.F. and Beck, J.N. (1995), "Why the Middle East fields may produce oil forever," *Offshore*, April, pp.56–62.

Manners, G. (1971), *The Geography of Energy*, London, Hutchinson.

Marchetti, C. (1978), *Energy Systems, the Broader Context*, Laxenburg, I.I.A.S.A.

Martinez, A.R. and McMichael, C.L. (1997), "Classification of petroleum reserves," *Proceedings of the 15th World Petroleum Congress*, Beijing.

Matveev, A.K. (1976), "Distribution and resources of world coal." *Proceedings of the 1st International Coal Exploration Symposium*, London, pp.77–88.

Maugers, L. (2003), "Time to debunk mythical links between oil and politics", *Oil and Gas Journal*, Vol.101.48, pp.18–28.

McCabe, P.J. (1998), "Energy resources: cornucopia or empty barrel." *AAPG Bulletin*, Vol.82, No.11, pp.110–34.

Meadows, D.H. et al. (1972), *Limits to Growth*, Washington, Potomas Association.

Meadows, D.H. et al. (2002), *Beyond the Limits: Confronting Global Collapse*, Washington DC, Chelsea Green.

Meyer, R.F. (1997), "World heavy crude resources," *Proceedings of the 15th World Petroleum Congress*, Beijing.

Meyer, R.F. and Olson, J.C. (Eds.) (1981), *Long-Term Energy Resources*, Boston, Pitman.

Mijamato, A. (1997), *Natural Gas in central Asia*, London, Royal Institute of International Affairs.

Mitchell, J. et al. (2001), *The New Economy of Oil*, London, Royal Institute of International Affairs.

Mohipour, M. et al. (2001), "Technology Advances in Worldwide Gas Pipelines," *Oil and Gas Journal*, Vol.99, No.48, pp.60–7.

Moody, J. M. (1970), 'Petroleum demands of future decades,' *AAPG Bulletin,* Vol. 50, No. 12.

Moritis, G. (2003), "CO_2 sequestration adds new dimension to oil and gas production," *Oil and Gas Journal*, Vol.101.9, pp.39–44.

Muir, W.L.G. (Ed.) (1976), "Coal Exploration," *Proceedings of the 1st International Coal Exploration Symposium*, San Francisco, Miller Freeman Publications.

National Energy Board (2000), *Canada's Oil Sands, Supply and Market Outlook*, Calgary.

National Research Council (2003), *Oil in the Sea III: Input and Effects*, Washington, DC, National Academies' Press.

Nehring, R. (1998), 'US reserves additions: innovation overpowering depletion,' *Oil and Gas Journal*, Vol.96, November 9, pp.87–89.

North, F. (1988), *Petroleum Geology*, Boston, Allen and Unwin.

Norwegian Ministry of Petroleum and Energy (2002), *Report on the Norwegian Continental Shelf*, Oslo.

Odell, P.R. (1964), *The Economic Geography of Oil*, London, G. Bell and Sons.

Odell, P.R. (1969), *Natural Gas in Western Europe: a Case Study in the Economic Geography of Resources*, Haarlem, de Erven F. Bohn.

Odell, P.R. (1971), *Oil and World Power*, London, Penguin Books, 1st edition and subsequently in seven more additions from 1972–1986.

Odell, P.R. (1972), "The geographic location component in oil and gas reserves' evaluation," *E.I.U. International oil Symposium*, London, pp.25–41.

Odell, P.R. (1973), "The future of oil: a rejoinder," *The Geographical Journal*, Vol.139, Part 3, pp.436–454.

Odell, P.R. (1974), *Energy; Needs and Resources*, Basingstoke, MacMillan Education.

Odell, P.R. (1975), *The Western European Energy Economy*, London, Athlone Press.

Odell, P.R. (1979), "World Energy in the 1980s: the significance of non-OPEC supplies." *Scottish Journal of Political Economy*, Vol.26, No.5, pp.215–255.

Odell, P.R. (1981), "Prospects for and problems of the development of oil and gas in developing countries," *National Resources Forum*, Vol.5. No.4, pp.317–26.

Odell, P.R. (1984), "The oil and gas resources of the Third World importing countries and their exploration potential," *Development Research and Policy Analysis Division of the United Nations*, New York.

Odell, P.R. (1985), "East-West differ on estimates of oil reserves," *Petroleum Economist*, Vol.52, pp.329–331.

Odell, P.R. (1988a),"The Western European gas market – current position and alternative prospects," *Energy Policy*, Vol.16, No.5, pp.480–93.

Odell, P. R. (1988b), 'Draining the world of energy,' in R. J. Johnson and P. J. Taylor (Eds.) *The World in Crisis,* 2nd edition, Oxford, Blackwell.

Odell, P. R. (1990), 'Continuing long-term hydrocarbons dominance of world energy markets: an economic and societal necessity,' *Proceedings of the World Renewable Energy Congress,* Oxford, Pergamon Press.

Odell, P.R. (1991), *Energy and the Environment,* Ditchley Conference Report no. D91/13, Oxford.

Odell, P.R. (1992), "Global and regional energy supplies – fictions and fallacies," *Energy Policy,* Vol.20, No.4, pp.284–96.

Odell, P. R. (1994a), 'Global energy markets: future supply potentials,' *Energy Exploration and Exploitation,* Vol. 2, No. 1, pp.39–72.

Odell, P.R. (1994b), "World resources, reserves and production." *Energy Journal Special Issue on the Changing Petroleum Market,* pp.89–114.

Odell, P. R. (1995), 'The cost of longer run gas supply to Europe,' *Proceedings of the SNS Energy Day 1995,* Stockholm, SNC Reprint Paper No. 25, pp.94–108.

Odell, P.R. (1997a), "The global oil industry: the location of oil production," *Regional Studies,* Vol.31, No.3, pp.309–20.

Odell, P. R. (1997b), 'Oil shock: a rejoinder,' *Energy World,* March, No. 247, pp.11–14.

Odell, P.R. (1998), "Oil and gas reserves: retrospect and prospect," *Energy Exploration and Exploitation,* Vol.16, No.2, pp.117–124.

Odell, P.R. (1998b), "Energy, resources and choices" in Pinder, D. (Ed.), *The New Europe; Economy, Society and Environment,* Chichester, J. Wiley and Sons Ltd.

Odell, P.R. (1999), "Dynamics of Energy Technologies and Global Change: a rejoinder," *Energy Policy,* Vol.27, No.12, pp.737–742.

Odell, P.R. (2000), "The global energy market in the long-term: the continuing dominance of affordable non-renewable resources," *Energy Exploration and Exploitation,* Vol.18, Nos. 2 and 3, pp.131–145.

Odell, P.R. (2001), *Oil and Gas: Crises and Controversies, 1961–2000,* Vol.1, *Global Issues,* Brentwood, Multi-Science Publishing.

Odell, P.R. (2002), *Oil and Gas: Crises and Controversies, 1961–2000.* Vol.2, *Europe's Entanglement,* Brentwood, Multi-Science Publishing.

Odell, P.R. and Rosing, K.E. (1980), *The Future of Oil: a Simulation Study of the Inter-relationship of Resources, Reserves and Use, 1980–2080,* London, Kogan Page, 1st edition. 2nd Edition in 1983.

Odell, P.R. and Vallenilla, L (1978), *The Pressures of Oil*, London, Harper and Row.

O.E.C.D. (2001), *Uranium; Resources, Production and Demand*, Paris, OECD.

Oil and Gas Journal (2002), "Focus on Turkmenistan," *Oil and Gas Journal*, Vol.100, Nos.41–44.

Oil and Gas Journal (2003a), "Gas in the US," *Oil and Gas Journal*, Vol.101, No.10, p.19 and Vol.101, No.11, p.23.

Oil and Gas Journal (2003b), "Future energy supply," *Oil and Gas Journal*, Vol.101, Nos.27–32.

Pachauri, R.K. (1988), "Energy and growth: beyond the myths and myopia," *The Energy Journal*, Vol.10, No.1, pp.1–20.

Pachin, J.C. (2003), "Coal has CO_2 capture opportunities," *AAPG Explorer*, Vol.24, No.8. pp.36–8.

Paik, K-W. (1995), *Gas and Oil in North East Asia: Policies and Prospects*, London, Royal Institute of International Affairs.

Paik, K-W. (2002), "Natural gas expansion in China," *Geopolitics of Energy*, Vol.24, No.5.

Paik, K-W. and Kim, D-K. (1995), "The Spratly islands' dispute with China's naval advance," *Geopolitics of Energy*, Vol.17, No.10, pp.5–10.

Park, D. (1979), *Oil and Gas in the COMECON Countries*, London, Kogan Page.

Parker, M. (1994), *The Politics of Coal's Decline; the Industry in Western Europe*, London, Royal Institute of International Affairs.

Peters, S. (2003), "Courting future resources conflict," *Energy Exploration and Exploitation*, Vol.21, No.1, pp.29–60.

Petford, N. and McCaffrey, K. J. W. (Eds.) (2003), *Hydrocarbons in Crystalline Rocks*, Bath, Geographical Society Publishing House.

Porfir'yev, V.B. (1974), "Inorganic origin of petroleum," *AAPG Bulletin*, Vol.58, No.1, pp.3–53.

Priddle, P. (1997), 'Natural gas: global security of supply,' *Proceedings of the 20th World Gas Conference*, Copenhagen.

Prin, R. et al. (2001), "Integrated Global System Model for Climate Policy Assessment," *Climate Change*, Vol.41, pp.469–546.

Quinn, A.C. (2000), "Long-term LNG contracts to opportunity markets." *Proceedings of the 17th World Petroleum Congress*, Rio de Janeiro, Vol.4, pp.185–192.

Radetzki, M. (Ed.) (1995), 'Coal in Europe: implications of dismantled subsidies,' *Energy Policy*, Vol. 23, No. 6, pp.481–560.

Radetzki, M. (1996), 'Fossil fuels will not run out,' *Journal of Mineral Policy,* Vol.12, No.2, pp.26–30.

Radetzki, M. (1999), "An economic analysis of climate policy," in Gerholm, T.R. (Ed.), *Climate Policy after Kyoto*, Brentwood, Multi-Science Publishing.

Randall, S.J. (2002), "Energy Security in the 21st Century," *Geopolitics of Energy*, Vol.24, No.4. April.

Rees, J. (1985), *National Resources: Allocation, Economics and Policy*, London, Methuen.

Rifkin, J. (2002), *The Hydrogen Economy*, London, Penguin Books.

Roberts, J. (1998), "Gas from the Caspian." *Geopolitics of Energy*, Vol.20, No.5, pp.1–3.

Rogner, H-H. (1996), *An assessment of world hydrocarbon resources*, Laxenburg, IIASA.

Rogner, H-H. (1997), *The Annual Review of Energy and Environment*, Laxenburg, IIASA, pp.217–262.

Rosa, R. (2003), "Climate change and oil depletion," *Energy Exploration and Exploitation*, Vol.21, No.1, pp.11–28.

Rowlands, I.H. (2000), "Beauty and the beast: BP's and Exxon's positions on global climate change," *Environment and Planning*, Vol.18, pp.339–354.

Ryan, J. (2003), "Hubbert's peak; déjà vu all over again," *IAEE Newsletter*, 2nd quarter, pp.9–12.

Sasanov, S. (2002), "The deep-water challenge," *World Petroleum Congress Report*, London, ISC Ltd, pp.120–7.

Schipper, L. and Meyers, S. (1992), *Energy Efficiency and Human Activity*, Cambridge, Cambridge University Press.

Schmoker, J. W. and Dyman, T. S. (1998), 'How perceptions have changed of world oil and gas reserves' *Oil and Gas Journal,* Vol.96, No.7, pp.77–79.

Schurr, S. and Netschert, B. (1977), *Energy in the American Economy, 1850–1975*, Baltimore, The Johns Hopkins University Press.

Seewald, J.S. (2003), "Organic/inorganic interactions in petroleum producing basins," *Nature*, Vol.426, No.6964, pp.327–32.

Shell IPC (1994), "Upstream Essentials," *Briefing Service*, No. 4.

Shell IPC (1995), "Energy in Profile," *Briefing Service*, No.2.

Shell IPC (1998a), *A Commitment to Sustainable Development: the Global Scenarios Project*, London.

Shell IPC (1998b), "Reflections on Kyoto," *World Economic Forum*, February.

Shell IPC (2001), *Energy Needs: Choices and Possibilities: Scenarios to 2050*, London.
Shell IPC (2003), *Inaugural presentation at the Shell Centre for Sustainability*, Rice University, Houston.
Shook, B. (1997), "Gas to liquids emerges from the fringe," *World Gas Intelligence*, December 19, pp.7–10.
Simon, J.L. and Kahn, H. (Eds.) (1984), *The Resourceful Earth*, Oxford, Blackwell.
Sinton, J. and Fridley, D. (2000), "Recent trends in China's energy consumption," *Energy Policy*, Vol.28, No.10.
Smil, V. (1998), 'Future of oil: trends and surprises' *OPEC Review* December, pp254–276
Smil, V. (2003), *Energy at the Crossroads: Global Perspectives and Uncertainties*, Cambridge, MIT Press.
Smil, V. and Knowland, W.E. (1980), *Energy in the Developing World: the Real Energy Crisis*, New York.
Smith, J. E. and McMichael, C. L. (1998), 'New reserves definitions approved by SPE and WPC,' *World Oil,* January, pp57–61.
Smith, N.J. and Robinson, G.H. (1997), "Technology pushes reserves crunch date back." *Oil and Gas Journal*, Vol.95, April 7, pp.43–50.
Spencer, J.A. (1990), "Advances in oil and gas reservoir appraisal," *Energy Exploration and Exploitation*, Vol.8, No.6, pp.191–202.
Stern, J.P. (1995), *The Russian Natural Gas Bubble*, London, Royal Institute of International Affairs.
Stinemetz. D. (2003), "Russian oil and gas sector rebound in full swing." *Oil and Gas Journal*, Vol.101, No.22, pp.20–30.
Styles, G. (2002), "Renewable energy is no free lunch," *Geopolitics of Energy*, Vol.24, No.10.
Styrikovich, M.A. (1977), "The long-range energy perspective." *Natural Resources Forum*, Vol.1, No.3, pp.252–63.
Tasch, J-H. (1980), *Coal Deposits: Origin, Evolution and Present Characteristics*, Sudbury, Mass., Tacsch Associates.
Thackeray, F. (1998), 'The future belongs to gas,' *Petroleum Review,* March, pp27–29.
Thackeray, F. (2002), "The promise of gas-to-liquids technology," 17th *World Petroleum Congress Report*, London, ISC Ltd., pp.176–83.
Thomas, R. and Ramberg, B. (Eds.) (1990), *Energy and Security in the Industrializing World*, Lexington, U.P. of Kentucky.
Toman, M.A. and Jemelkova, B. (2003), "Energy and economic development: an assessment of the state of knowledge," *The Energy Journal*, Vol.24, No.4, pp.93–11.

Torp, T.A. (2001), "Carbon sequestration: a case study," *17th World Petroleum Congress Report*, London, ICS Ltd., pp.156–9.

United Nations (2000), *Trade Agreements, Peroleum and Energy Policies*, UNCTAD, Geneva.

United Nations (2001a), *Sustainable Energy, Shifting towards a Development Path*, New York, UN Series No.38.

United Nations (2001b), *World Population Prospects; the 2000 Revision*, New York, Population Division.

United Nations (2002), *World Energy Assessment*, New York, Development Programme.

United States Congress, Office of Technology Assessment (1992), *Fuelling Development: Energy Technologies for Developing Countries*, Washington DC, Government Printing Office.

United States Geological Survey (1997), *World Energy Resources*, Washington DC, Government Printing Office.

United States Geological Survey (2000), *World Petroleum Assessment*, Reston, Virginia, Government Printing Office.

United States Geological Survey (2001), *Natural Gas Hydrates: Vast Resources, Uncertain Future*, Reston, Government Printing Office.

Van de Vate, J. (1997), "Comparison of energy sources in full chain emissions of GHG," *Energy Policy*, Vol.25, No.1, pp.1–6.

Van Vuuren, D. et al. (2003), "Energy and emissions scenarios for China in the 21st century," *Energy Policy*, Vol.31, No.4, pp.369–88.

Verbicky, E. (1998), "Oil sands: a growing and viable alternative to conventional oil," *Petroleum Economist*, Vol.65, No.1, pp.21–3.

Vrolijk, C. (2002), *Climate Change and Power*, London, Royal Institute of International Affairs.

Vyakhirev, R.I. (1997), "Natural gas in Russia: potential for the 21st Century," *Proceedings of the 15th World Petroleum Congress*, Forum 10, Beijing.

Warman, H.R. (1972), "The future of oil," *Geographical Journal*, Vol.138, No.3, pp.287–97.

Wälde, T.W. and Gunst, A.J. (2002), "International energy trade and access to energy networks," *Journal of World Trade*, Vol.36, No.2, pp.191–218.

Wasserstrom, R. and Reider, S. (2003), "Oil project lending faces new environmental litmus test (2003)," *Oil and Gas Journal*, Vol.101, No.39, pp.35–6.

Watkins, G.C. (2003), *Characteristics of North Sea Oil Reserves Appreciation*, Center for Energy Policy Research, Massachusetts Institute of Technology (MIT-IEEPR 2000-08 WP).

Wilkins, G. (2002), *Technology Transfer for Renewable Energy,* London, Royal Institute of International Affairs.
Williams, B. (2003), "Heavy hydrocarbons to play key role in future energy supply," *Oil and Gas Journal,* Vol.101, No.29, pp.20–7.
Williams, R.H. (1998), "A technological strategy for making fossil fuels environment and climate friendly," *World Energy Council Review,* September, pp.59–67.
Williamson, H.F. et al. (1963), *The American Petroleum Industry, 1859–1959,* Evanston, North Western University Press.
World Energy Council (1993), *Energy for Tomorrow's World,* London, Kogan Page.
World Energy Council (1998), *Global Transport and Energy Development,* London, World Energy Council.
World Energy Council (1999), *The Challenge of Rural Energy Poverty in Developing Countries,* London, World Energy Council.
World Energy Council (2000), *Energy for Tomorrow's World – Acting Now!* London, World Energy Council.
World Energy Council (2001), *Living in One World,* London, World Energy Council.
World Petroleum Congress (2002a), "New hydrocarbons provinces of the 21st century," *Proceedings,* Vol.2, Forum 2, pp.87–176.
World Petroleum Congress (2002b), "Exploration and production in environmentally sensitive areas," *Proceedings,* Vol.2, Forum 6.
World Petroleum Congress (2002c), "Natural gas: clean energy for half-a-century," *Proceedings,* Vol.4. Forum 14.
Wybrew-Bond, I. and Stern, J. (2002), *Natural Gas in Asia,* Oxford, Oxford University Press.
Xu Yong Chang and Shen Ping (1996), "Natural gas origins in China," *AAPG Bulletin,* Vol.80, No.10, pp.1601–14.
Yang Jingmin et al. (1997), "Analysis of world oil supply and demand and the development trend of the Chinese petroleum industry," *Proceedings of the 15th World Petroleum Congress,* Beijing.
Zarilli, S. (2003), "Domestic taxation of energy products and multilateral trade rules," *Journal of World Trade,* vol.37, April, pp.359–394.
Zimmerman, E.W. (1951), *World Resources and Industries,* New York, Harper and Row.

Statistical Sources
BP (annually 1961–1982), *Statistical Review of the World Oil Industry,* London

BP (annually 1983–2003), *Statistical Review of World Energy,* London
BP Exploration (1996, 7 and 8), *Supplementary Gas Data,* London.
BP (annually 1990–5), *Statistical Review of World Gas,* London.
Cedigaz (annually 1988–2003), *Natural Gas in the World*, Paris, I.F.P.
Energy Information Administration (annually), *International Energy Outlook*, Washington D.C.
International Energy Agency (annually), *Oil and Gas Information*, Paris, OECD.
International Energy Agency (quarterly), *Oil Market Report*, Paris, OECD.
International Energy Agency (2003), *Renewables Information*, Paris, OECD.
United Nations (annually), *Energy Statistics Year Book*, New York.
United States Geological Service (2000), *World Petroleum Assessment*, Washington, D.C.
World Energy Council (1998), *Survey of Energy Resources*, London.
World Renewable Energy Network (2001), *Renewable Energy*, Reading, Sovereign Publishing.

Relevant Journals
AAPG Explorer, Houston, US.
Bulletin of the American Association of Petroleum Geologists, Houston, US.
Energy Economist, London, UK.
Energy Exploration and Exploitation, Brentwood, UK.
Energy in Europe, European Commission, Brussels.
Energy Policy, Oxford, UK.
Energy World, London, UK.
Gas Matters, London, UK.
Geopolitics of Energy, Calgary, Canada.
Greenhouse Issues (IEA), Cheltenham, UK.
Journal of Petroleum Geology, London, UK.
Middle East Economic Survey, Cyprus.
Nature, London, UK.
Natural Resources Forum, New York, US.
Oil and Gas Journal, Houston, US.
OAPEC Bulletin, Kuwait.
OPEC Review, Vienna, Austria.
Oxford Energy Forum, Oxford, UK.
Petroleum Economist, London, UK.
The Energy Journal, Pittsburgh, US.
World Oil, Houston, US.

Index

A

Abiogenic theory of oil and gas' origins	112–121
Access to domestic electricity	5–6
Accidents	56
Africa	20, 21, 29, 74, 76, 83, 85, 91–2
north	87, 91
west	84
Alaska	57
Alberta	51, 58, 65, 102
oil sands	58, 65, 102
Algeria	58, 86, 91, 92, 113
Alternative energy resources	xi, 10, 12
Alternative theory of the origin of oil and gas	xxiv, 66, 112, 121
All-Union Conferences on Petroleum and Petroleum Geology	116
America	xvii
Central	74, 76, 83, 85
North	19, 20, 21, 39, 56, 58, 62, 63, 72, 74–6, 83–4, 86, 91, 92
South	24, 29, 62, 63, 74, 76, 82, 83, 85
American Association of Petroleum geologists (AAPG)	121
Annihilation of space	4
Annual additions to oil reserves	43
Anthropogenically derived CO_2 emissions	xi, 80

Antarctica	59, 75, 92
Appreciation of reserves	xxii, 35–45, 73
Arctic	58
located oilfields	57
Argentina	85
Asia(n)	xvii, 27–29, 117
central	89
dependence on energy imports	64
– Pacific	27–29, 63, 65, 74, 76, 82, 83, 93–4
region oil reserves	61–3, 65
relations with the Middle East	64, 90
Athabasca	51
Australia	24, 28, 29, 30, 94
Australasia	20, 21
Availability of carbon fuels	xvii, xix, 1, 13, 14, 19–22, 35, 39, 46, 50, 52, 72, 75, 77, 101, 112, 114, 117, 120–1
Availability of renewable energy	xvii, 23
'Away from Middle East' policies	62
Azerbaijan	41, 57, 87, 89, 117

B

Backdating of oil reserves' discoveries	43
Bangladesh	94
Barents Sea	86–7
Belarus	xvi
Belgium	xvi, 25, 26
Biogenic carbon energy	111–12, 113, 116, 117, 121
Biogenic origin of coal, oil and natural gas	xi, 116
Biomass	xiii, xvi
Bitumen	51
Bolivia	85
Boundary claims	95
Brazil	25, 51, 85
British Petroleum (BP)	10, 35, 63
Brunei	64, 94

C

Calgary	iv
Cambodia	64
Cameroon	57
Canada	xxi, 24, 51, 58, 61–63, 84, 105, 113
heavy oil & tar sands	58, 102
Canadian Energy Research Institute	iv
Capital costs/investments	103–5
Carbon energy(ies)	xi–xiv, xvi–xxiii, 8, 10, 12, 13, 19–22, 59, 79, 80, 91, 101, 102–6, 111, 114
Carbon fuels	xi–xviii, xx, 8, 10, 12–16, 59, 60, 71, 79, 80, 89, 101, 102, 104, 108, 111
Caspian Basin	63, 117
Caspian Sea	87
Caucasus	113
Causal link between carbon fuels' use and CO_2 emissions	xvii
Cement industry	5
Centrally planned economics	xiv, xvii, 4, 8
Chad	57
Cheap energy	4
Chemical genesis of hydro-carbon molecules	114–5
Chemical thermodynamic environments of the earth's crust	xxiv, 112
Chile	xvi, 85
China	8, 19, 24, 30, 51, 61, 87, 93, 94, 113
coal industry	28
energy use	61, 64
exports	28
railway system	28
state owned offshore oil and gas	64
City regions	29
Clean coal technology	23, 107, 112
Climate change	xii, xvii, xviii, 60

Club of Rome	1
CO$_2$ emissions	xi, xiii–xvi, 9, 23, 29, 31, 71, 80, 81, 94, 107
Coal:	
as the fuel of the 19th century	xxiii
bearing basins	23
capital investments	103
cumulative production	22
demand for	xix, 1, 14, 20, 26
dependent countries	29–30
depletion	xii, 21
environment constraints	22–3
environmental taxes	106
expansion prospects	19–32
exploration/exploitation	xvi, 13, 14, 19–23, 25, 28, 29, 31, 79, 103, 107
Exploration Symposium	20
exporters	24, 26, 28
gasification plants	30, 31
geographical concentrated pattern of production	xx
importers	23, 25–29
lack of acceptability	xx
markets	xix, 22, 30
peak production	28
prices	25, 29, 104, 106
production potential	23, 28
proven reserves	xix, 21, 22, 24, 28, 102
regional supply	26
reserves	xix, 19–26, 28, 29, 102, 104
reserves potential	24, 28
resources	xii, xx, 14, 19–32, 60, 79, 107
scarcity	xii, 21, 22
share of energy use	xx, 22, 23
to oil products conversion plants	31
underground gasification	30
uneconomic production	25, 26
use	19–22, 27–29
Coal-bed methane	xxiii, 75
– fired electricity generation	27
Coal-measures gas	76, 84

"Cold" eruptive processes	115
Cold War	120
Colombia	24, 26
Combined-cycle generation	8, 31, 94
Competition between energy sources	xxii, 12, 22, 26, 27, 42, 53, 102, 107
Communist regimes	9, 57
Competition for markets	xxii, 23, 27, 42, 95
Compressed natural gas	96
Constraints on:	
fuel use	xiv, 56
oil supplies	xxi, 31
production	1, 22, 56
Continental shelf(ves)	13, 86
Continental slopes	13
Conventional gas	xxiii, 73–77, 80, 84–7, 89, 91, 93, 96, 105–8
Conventional oil	xxi, xxii, 45, 50–2, 59, 61, 63, 65, 89, 102, 104, 105, 107
Cost differentials between carbon fuels and renewables	14
Costs of CO_2 sequestration	xiv, 80
Crude oil	5, 6, 20, 31, 56, 103, 104, 116
Cuba	8
Cumulative production of:	
gas	xxiii, 77, 89
oil	10, 12, 44
Czech Republic	xvi, 24, 26, 30

D

Deep oceans	58, 84
gas hydrates	13
offshore oilfields	59
Deep water, oil and gas reserves	13, 59
Demand for energy	xi, xiii–xiv, xvii, xix–xx, 8, 9–16, 64, 79, 92, 102, 107, 108, 111–113
Demand side competition	14

Demographic trends	xv
Denmark	8, 25, 26, 30
Depletion curve for:	
conventional gas	74, 75, 78
conventional oil	48, 49, 52, 59
non-conventional gas	77, 78
non-conventional oil	49, 52
Depletion of natural resources	xii
Developing countries	xv, 8, 46, 57, 81
Diesel oil	31
Dnieper/Donetz basin	13, 113, 117, 119
Drilling and production facilities	57
Dutch Frisian Islands	57

E

East Atlantic margin	58–9
East Siberia	64, 89
Economic:	
difficulties	xii
geographical considerations	xxii
growth	xii, xiv, 8, 9, 64, 93
rent	105
structures	xiv
Economic and social desirability of increased carbon energy use	xvii
Economies of scale	14, 107
Egypt	91
Electricity:	
generation	23, 27
production	xiii, 7, 23
systems	xiii
Electrification	4, 27
Emissions	xi, xiv–xviii, 9, 23, 26, 60, 71, 80, 82, 96, 107
controls	xiii, xiv, 25, 29, 31, 32, 60, 81, 94
"End of history"	36, 46
Energy:	
and economic development	1, 4

careless and wasteful systems	7
competition between hydro-carbons and renewables	xxiii
conservation	7, 9
demand/use	xi, xii, xix, xx, xxi, 1, 8, 9, 16, 35
demand expectations	75
development	30–2, 65, 94–6, 107–8
economics	96, 107
efficiency	xiv, xvii, xviii, 6–9, 79
intensive countries	xvii, 29
intensive use	5, 60, 92
intensity of economic growth	7, 8
markets	xiv, xvii, 92
policy making	73
processing	xii, 31
resources	xi, xii, xiv
supplies	xi, xvii, xx, 9, 53
supplies by source	xx
supply gap	20, 79
supply requirements	xix, 9
Energy World	114
Energy-plexes	31
Enhanced oil recovery	39, 58
Environment:	
concerns	xix, xx, 9, 13, 14
constraints	14, 22–3, 32, 56–60, 59, 79, 82
lobby	58, 59
taxes	79
Environmentally-friendly:	
attitudes	xviii, 56, 58, 97, 107, 112
policies	59
Erasmus University	iv, 36, 48
Centre for International Studies	36
Ethical justification for carbon energy use	xvi
Eurasia	91
gas markets	89
Europe(an):	20, 21, 74, 76, 83, 88
central	26–7
coal imports	26–7
coal policies	26
constraints on gas production and use	31, 82, 90, 112

Europe(an) (*continued*)
    Eastern   xiv, 4, 8, 9
    emerging continental supply system   87
    emerging energy economy   27
    gas' contribution to the economy   86
    gas imports   90
    markets   xxii, 26, 62, 82, 85–7, 92
    North-West   59, 72
    Western   6, 8, 26–7, 72
    Commission   xv, xviii, 32, 60, 72, 82
    Court of Justice   59
    Union (EU)   26, 59
Expectations for future requirements of
    carbon energy   9–12
Exploitation of coal, oil and gas   xii, xvi, 14
Exploration costs   57, 73, 103, 105
Exxon-Mobil Corporation   63

F

First oil price shock   7, 36
Final consumer of energy   6
Finding and development costs   102–103
Finite resources of oil and gas   xxiv, 112
'Flat-earthers' (theory)   46, 48
Forecasts of:
    energy use   10–12
    oil and gas use   xvii, xix, 10–12, 36
Foreign exchange costs of imported oil   8
Former Soviet Union (FSU)   xiv, xxiv, 4, 8, 19, 20, 21, 35, 39, 51, 72, 73, 74, 76, 82, 83, 86–9, 90, 15–8, 120–1
    competition with the Middle East   63, 90
    Republics   41, 87, 89, 90
'Fossil' fuels   xi, 66, 117
Fragile environments   xxiv, 56, 57
Fragility of hypothesis of oil and gas
    biological origins   112
France   xvi, 25, 61, 62

G

Gas (see natural gas)	
Gas hydrates	xxiii, 13, 75, 77, 79, 85, 89, 90
Gasoline	31
Gas to liquids (GTL)	95–6
Gas Resources Corporation	114
Genesis of heavier hydro-carbons	115
Geological Congress	19
Geo-pressured gas	77
Geopolitical:	
constraints	xi, xxi
importance	xxii
Geopolitics	63, 94, 114
Geothermal power	114
Germany	xi, 8, 24, 25, 29, 30, 61, 62
Giant fields	xxiv, 85, 113, 117
Global warming	xi, xii, xvi, xvii, 9, 23
Greece	29, 30
Green energy	xiii
Greenhouse gas emissions	xvii, 82, 96
Greenpeace	xvi, 58
Groningen gas field	85
Great rift valley	92
Guerrilla actions against oil installations	57
Gulf of Mexico	xxii, xxiv, 50, 84, 113, 121
Gulf War	56

H

Habitats of oil	xxi, xxiv, 50, 121
Halbouty, M.	121
Heavy oils	50, 51, 58, 102
Hedberg Research Conference	121
High-pressure polymorphs	115
Horizontal drilling	39, 57
Household-goods industry	45
Houston	114
Hubbert's hypothesis	44

Hungary	30
Hydrocarbons	xxi, 14, 22, 29, 74, 75, 86, 89, 94
as renewable resources	xxiv, 112–122
supply	xx, xxi, 53, 76, 77, 81
Hydrogen	31, 115
production from natural gas	xviii
use	60, 96

I

Incremental demand for energy	xiii
India	24, 25, 28, 30, 51, 61, 87, 90, 93–4, 113
Indigenous sources of carbon energy	xvi, xvii, 6, 25, 26, 58, 86
Indonesia	24, 58, 64, 93, 94
Industrialisation:	
policies	5
processes	4
Infant industry(ies)	xiii
Initial reserves of oil	43, 51
Inorganic origins of oil and gas	xxiii, xxiv, 112–14, 120
Insulation of buildings	7
Inter-continental transport	62
Intergovernmental Panel on Climate Change (IPCC)	xviii
International:	
Energy Agency (IEA)	iv, xv, xiv–xvi, xviii, 8, 23, 31, 62, 80, 81, 85, 93, 103, 104
Gas Union (IGU)	72, 74, 75, 83
Institute for Applied Systems Analysis (IIASA)	iv, 75, 77, 120
movements of oil	61–2
oil companies	23, 41, 46, 63, 64
oil industry	44, 59, 63
trade	93, 101
Interquartile price range for oil	104
Intra-regional trade	62
Investment	12, 23, 39, 42, 57, 103
capital	8

in coal	23, 103
in oil	51, 63, 65, 103, 105, 106
in natural gas	53, 73, 77, 103, 106
IPCC	xviii
Italy	25, 61
Iran	41, 58, 61, 86, 91
Iraq	41, 57, 58, 61
Ireland	30
Iron and steel industry	5

J

Japan	xvii, 6, 25, 27, 28, 61, 64, 93, 94
Joint Institute of the Physics of the Earth	114
Joint ventures	41, 63
Journey-to-work	4

K

Kansas	113
Kazakhstan	24, 30, 41, 87, 89, 93
Kenney, J.F.	112, 114, 117, 118, 120
Komi	113
Korea, South	xvii, 27, 28, 29, 93, 94
Kudryavtsev, Prof. N.	115
Kuwait	41, 56, 58
Kyoto Treaty (Protocol)	xiii, xiv, xvi, 29, 31, 81

L

Landless labourers	5
Latin America	20, 21, 24, 29, 82
Latin American preference for natural gas	29
Leakages from gas pipelines	82
Leases for exploration	63
Leisure time	4

Liberal trading regimes	6
Liberalisation of markets	82, 83
Lifestyles	4, 5
Lignite	28
Liquefied natural gas (LNG)	80, 86, 87, 91, 93, 94
– in Asia/Pacific region	93
– liquefaction	80, 91, 103
– re-gasification terminals	80, 103
– United States imports	84
Living standards	xvi, 4, 5
London	19, 62
School of Economics	iv
Louisiana	50
Low-cost energy	xii, xv, 1, 102, 104

M

Madagascar	51
Malaysia	64, 93, 94
Managerial achievements	xii
Manufacturing industry	5
Marine environment	58, 59, 80
Maritime routes	56
Mechanisation of households	4
Mediterranean coast/littoral	57, 91
Mediterranean pipelines	57, 91
'Mega-majors'	63
Metal fabrication	5
Methane (see also natural gas)	75, 84, 115, 118
as a greenhouse gas	82
emissions	82
Mexico	xvii, xxiv, 50, 61, 62, 121
gas prospects	xxii, 84, 113
Middle East	7, 29, 39, 45, 62, 63, 71, 74, 76, 83, 94, 105
contrasts with the FSU	89–90
gas reserves	90
links with SE Asia	64–5, 90, 93
LNG	87, 89–91, 94

oil production	xxiv
oil reserves	45–6, 60–1, 113
Moscow State University	21
Motor vehicles	4, 7, 31, 32, 60, 96
Motorisation of societies	xv
Mozambique	92

N

Namibia	92
Nationalisation of oil	39, 46, 62
Nationally protected environments	57
Natural gas:	
additional reserves	xxiii, 73–4, 85, 87, 90, 93
as the fuel of the 21st century	xxii–xxiii, 71–97
as a renewable resource	xxiii–xxiv, 111–121
availability	14, 72, 75, 76, 77, 84, 101–2, 121
demand/use	xix, xxii, 1, 12, 14, 71–3, 75, 77, 79, 82, 84–87, 90–2, 94, 96
environmentally friendly characteristics	xviii, 80–2
from fractured shales	xxiii, 75, 84, 85
industry	xii, xiii, xvii, xxi, 72, 73, 82, 83, 85, 86, 92, 106, 107
in transport(ation) use	29, 96
markets	xiii, xxii, 53, 71, 73, 80, 82–94, 106
pipelines	31, 80, 82, 86, 87, 90–4
prices	83, 84, 86, 104, 106, 107
production	xxii, xxiii, 51, 71–97, 105, 107, 108, 114
regional demand variations	82
reserves	xxii, xxiii, xxiv, 14, 71–7, 83–88, 90–97, 113, 114
reserves to production ratios	71, 72, 73, 87, 93, 114
resources	xii, xxii, xxiii, xxiv, 14, 74, 75, 79, 85, 86, 87, 89, 90 ,93, 94, 102, 112, 114, 121
rich provinces	xxii, 72

Natural gas (*continued*)
    transmission lines   85, 86, 92
    use technology   23, 94, 96
Nature of resources   13, 121
Netherlands   25, 26, 72, 86
'Neutral stuff'   13
New York   62
Niche markets for renewables   xv
Nigeria   56, 91, 92
    LNG exports   84, 91
Non-carbon energies (see renewable energies)
Non-conventional:
    oil and gas   xxii, xxiii, 49–53, 63, 75–79, 84, 85, 87, 89, 90, 96, 102, 105, 106, 107, 108, 112, 114
    oil production costs   51, 107
    production   xxi, xxiii, 51–3
    reserves   xxi, 52, 63, 65
Non-depleting fields   118
North America   19–21, 39, 56, 59, 72, 74–76, 83
    gas markets   83, 84, 86, 91, 92
North Cape   86
North Sea   xxii, 39, 50, 63, 86, 87, 113
Norway   61, 62, 72, 73, 86
Nuclear electricity/power   xiii, xvi, 12, 20, 27

O

OECD   xv–xvii
Offshore oil and gas   50, 58, 59, 64, 72, 73, 85, 89, 90, 92, 96, 102
Oil:
    as the fuel of the 20th century   xii
    as a renewable resource   111–121
    availability   xix, 35, 39, 43, 46, 52, 102, 112, 120, 121
    companies   5, 7, 23, 35, 39, 41, 46, 51, 58, 59, 63, 64, 65, 104, 121
    consuming countries   61, 64

crises	21, 27, 41, 62
cumulative use	xii, xix, 10, 12, 36, 44
declining importance	xxi–xxii
demand/use	xix, xxi, 1, 10, 14, 31, 36, 39, 42, 45, 46, 52, 60, 61, 64, 82, 104
depletion	xii, 45, 48, 52, 59
discoveries	xxii, 36, 43, 45, 113
exporting countries	7, 57, 65, 105
geology/geologists	xxiv, 10, 12, 13, 35, 39, 48, 50, 59, 112, 115, 116, 117, 120, 121
importing countries	5, 6, 8, 62, 64, 102
initial reserves	51
long-term future	xxi, 112, 121
low-cost	xv, 5
markets	41, 50, 53, 62, 105
peak production	xxi, 48, 52, 60, 121
pessimism on its future	35, 42
policy making	35, 41, 58
political significance	xxii
producing countries	5, 61
production	xvii, xxi, xxii, 6, 10, 12, 35, 36, 39, 40, 43, 44, 50, 51
prices	7, 42, 43, 51, 60, 63, 102, 104, 105
prospects	xxiv, 12, 35, 36, 43, 45, 46, 58, 59, 60–5, 114, 120, 121
refining	50
reserves (proven)	xxi, xxii, xxiv, 13, 14, 16, 35, 36, 39, 41–8, 51, 52, 58–63, 65, 71–3, 104, 113
reserves additions	12
reserves appreciation	35–41, 44, 45
resources	5, 14, 39, 42, 43, 45, 48, 52, 59, 62–5, 72, 72, 112, 114, 120, 121
scarcity	xii, 20, 21, 35, 36, 41, 44, 45, 46
share of cumulative hydrocarbons production	xxi, 53
supply/ies	xxi, xxii, xxiv, 5, 9, 14, 16, 35, 36, 39, 41, 42, 43, 46, 50, 51, 52, 53, 56, 58–63, 102, 104, 105, 112, 114, 121
transport	56, 57, 62, 103

Oil and gas as renewable resources	xxiii–xxiv, 111–121
Oil and gas plays	xxiv, 113
Oil from coal plants	31
Oilfields' discovery process	35–41, 45
Oman	91, 94
Operating costs of production	103
"Ordered" markets	104
Organic theory of coal, oil and gas' origin	xxiii, xxiv, 111, 112, 113, 121
Orimulsion	51
Orinoco oil belt	51
Organisation of European Co-operation and Development (OECD)	xv, xvi, xvii
Organisation of Petroleum Exporting Countries (OPEC)	58, 102–4

P

Pacific	63, 64, 65, 82, 93–4
Papua New Guinea	94
Paraguay	xvii
Paris	iv
Peak oil production	xvii, xxi, xxii, 35, 36, 41, 45, 48, 52, 60, 65, 107, 121
Peasants	5
Per capita use of energy	xv, xvii, 4, 8
Periods of energy price rises	107
Peru	85
Pessimism over oil availability	10
Petrochemical products	4
Petroleum:	
and the mainstream of modern science	117
below the methane clathrate zones	118
exploration neither mature nor declining	118
from crystalline basement	117
from impact structures	117
from non-sedimentary basins	117
from volcanic structures	117
geology	112, 120

industry's adolescence	118
potential of riftogenic suture zones	118
products	5, 31, 50
Petroliferous provinces	xxiv
Philippines	xvii, 64
Pipelines	27, 31, 57, 80, 82, 86, 87, 90–4
Plentiful energy resources	xii, 45, 102, 105
Poland	24, 26, 30
Political changes and difficulties	xii, 90, 92, 94, 105
Population	xiv, xv, xvii, xx, 4, 5, 8, 14, 93, 107
post-2050 slower growth rate	14
Populist protests against energy prices and taxes	xiv
Porfir'yev, V.B.	113, 120
Power generation	23, 72, 92, 94, 106
Pre-Cambrian crystalline basement	xxiv, 113
President Bush	58
President Kennedy	120
President Khrushchev	120
Pressurised hydrocarbons from the mantle	116, 118
Price(s):	
decline in real terms	6, 8
inelastic supply side curve	36, 72
long-term supply	46, 104, 111
of carbon energy	xviii, 77, 101, 102, 102, 104, 108
of internationally traded crude oils	104
of oil	5, 42, 44, 51, 56, 62, 91, 104, 105, 106
preferred	104
upward pressures	7, 12, 46, 63, 105
Primordial material	115
Private transport	4
Process of economic development	xv, 29
Production costs of energy	xii, 43, 101–8
Production curves for gas	xxii, 74, 76
Production curves for oil	xxi, 52
Profit margins	103
Protection for indigenous production	101
Public opinion on energy issues	xiv
Public transport	4

Q

Qatar	91, 94

R

Radical policies on energy markets	xiv
Rates of increase in use of (demand for) energy	xix, 1, 5, 7, 9, 107, 111
Rates of increase in recovery of oil	39, 46
Reduction targets for CO_2 emissions	xiv, 81
Re-globalisation of science	121
Relationship between energy use and economic development	1, 4
Remaining undeveloped oil frontiers	59
Renewable energy resources	xiii, xiv, xv, 27, 29, 66, 79, 107
demand for	xiv, iii
falling real costs	xiv, 108
production plants	xiii
prospects	xvi, 79
subsidies for producers	xiii, 79, 106
supply	107
Repleting oil and gas fields	xxiv, 113, 114
Reserves to production ratio	xxii, xxiv, 12, 22, 28, 42, 71, 72, 73, 87, 93, 114
Reservoirs of oil/gas	13, 39, 46, 50, 115, 117, 118, 121
Resource abundance	71–5
Resource wars	xii
Returns on investment	101, 106
Revenues from oil	7, 58
Revolution of rising expectations	4
Rigorous analytical theory of origins of oil and gas	115, 116
Royal Dutch (see Shell)	
Rural living	5
Russia(n):	8, 19, 24, 26, 30, 41, 56, 57, 61, 62, 64, 65, 86, 90, 91, 93
Academy of Science	114
oil companies	63

pipelines	87, 89
reserves of gas	73, 86, 87
Ukrainian theory of oil's origin	xiv, 65–6, 112–118

S

Sakhalin	89
Saudi Arabia(n):	41, 61
light crude	5, 6
Scarcity of energy resources	xi, xii, xix, 1, 20, 21, 35, 36, 41, 44, 46, 72, 105
Second law of thermodynamics	116
Second oil price shock	7, 101
Sequestration of CO_2	xiv, xv, xviii, 31, 80, 96, 107, 112
Seismic methodologies	39
Service sector	5
Sewer gas	115
Shell International Petroleum Company	12, 36, 39, 48, 51, 56, 63, 71
Shipping	115
Siberia	63, 64, 89, 113, 117
Singapore	62, 94
Slovakia	25, 26
Societal changes	xiv, xvi, 5, 7
Societal structures	xiv
Solar power	xiii, xvi, xix, 10, 23
South Africa	24, 30, 92
coal	26, 32
South China Sea (dispute)	64, 94
South East Asia	61, 62, 87, 89, 90, 93, 94
coal imports	27–9
oil imports	62
South Korea	xvii, 25, 27, 30, 29, 61, 94
Soviet Union	xiv, xxiv, 4, 9, 19, 39, 51, 63, 72, 73, 82, 87–9, 115–18, 120, 121
Spain	25, 30
Spatial mobility	xv
Spratly Islands dispute	64, 94
Stabilisation of CO_2 emissions	xiv
Standards of comfort in homes	4

Standards of social welfare	xv
State oil entities	41, 64, 105, 106
Styrikovich, Prof. M.A.	46, 65, 120, 121
Sub-bituminous coal	28
Sub-Saharan Africa	xvii
Substitution of coal and oil	xii, 26, 27, 80
Subterranean micro-biological habitat	121
Suburbanisation	4
Sufficiency of energy	xvi
Super-normal profits	104
Supply costs	102–4
Supply price curve	7, 45, 102–5
Sustainability	xvi
Swamp gas	115
Sweden	xvi
Switch to renewable energies	xi, xx

T

Taiwan	25, 29, 30, 94
Tankers	56, 80, 87
Tatarstan	117
Taxes:	
on carbon fuels	60, 79, 105, 106
on oil production	104
Technological:	
achievements	xii, 5
developments	xiv, xxii, 7, 42, 46, 84, 102
Texas	50, 113
Textile industry	5
Thailand	25, 94
Thermodynamics, 2nd law of	xxiv, 112, 114–117
Third World	9
Tidal power	xiii
Tight formation gas	xxiii, 75, 85
Toronto	19
Trade	62, 89, 91, 93, 101, 104
in energy	xi
Transmission lines	85, 86, 92

Trans-Saharan pipeline	92
Transport	4, 7, 28, 60, 89, 91, 96, 101
inter-continental	62
of oil	56–60, 62
sector economics	xiv, 60, 96
Turkey	xvii, 30, 57, 89
Turkmenistan	41, 86, 87, 89, 93

U

Ukraine	xiv, 24, 30, 41, 65–6, 87, 113
Ultimate conventional oil depletion	45–8
Ultimate world oil resource base	42, 59, 120
Ultra-deep gas	xxiii
Unexplored habitats of oil and gas	13, 46, 73
United Arab Emirates	58, 94
United Kingdom (UK)	8, 24–26, 30, 58, 59, 72
United Nations (UN)	14, 50
Institute for Training and Research (UNITAR)	50
United States (US)	5, 24, 26, 28, 35, 50, 51, 57, 58, 61, 62, 64, 72, 76, 77, 82, 83, 84, 113, 115
gas imports	84, 91, 92
Geological Survey (USGS)	48, 74
natural gas industry	6, 72
non-conventional oil and gas	84
oil import quotas	102
Urals	87, 113
Urbanisation	xv, 5
Uruguay	xvii
Uzbekistan	87, 89

V

Vehicle production	5
Venezuela	xxi, 51, 58, 61, 63, 85, 91, 105, 113

Vienna	iv, 75, 120, 121
Vietnam	8, 9, 64, 113
Volga-Urals	113

W

Wasteful consumption (use) of energy	6
Wave power	xiii
Western 'oil-men'	121
Western Pacific rim	26, 27–9
Wind power	23
Windmills	xiii
World:	
Bank	57, 103
Energy Assessment (USGS)	14, 48, 59, 73, 74, 75, 77, 86, 91
geology	xxiv, 13
population	xiv, xv, xvii, xx, 8, 14, 107
Petroleum Congress (WPC)	xviii, 63, 71, 72, 80
War II	1, 35
Wyoming	113
Wytch Farm oilfield	57

Z

Zaire	51